Unraveling Reality:
Exploring the Mandela Effect Phenomenon

Kimberly Wylie

CYPRESS CANYON PUBLISHING

Copyright © 2024 by Cypress Canyon Publishing

All rights reserved.

No part of this book may be reproduced in any form or by any electronic or mechanical means, including information storage and retrieval systems, without written permission from the author, except for the use of brief quotations in a book review.

For information contact:

http://www.cypresscanyonbooks.com

ISBN: 9798334856912

Dedication

To everyone who is both disturbed and intrigued by these glitches in our reality, this book is for you. May your curiosity lead you to deeper understanding and wonder.

TABLE OF CONTENTS

Introduction 7

CHAPTER 1: What is the Mandela Effect?............................... 9
    Cognitive Biases and the Mandela Effect................................ 10

CHAPTER 2: Understanding Memory and Perception........... 13
    Memory Formation and Retrieval 14
    Factors Influencing Memory Accuracy 15
    Role of Perception in Shaping Our Reality 16

CHAPTER 3: Famous Mandela Effects................................... 19
    Nelson Mandela's Death 19
    Berenstein Bears vs Bearenstain Bears 22
    Where the Heck is New Zealand?.. 26
    Monopoly Man's Missing Monocle 32
    Sinbad as *Shazaam!*.. 39
    The Fruit of the Loom Cornucopia 42
    Sleeping Beauty 47
    C-3PO's Leg.. 49
    "Luke, I am Your Father." 57
    King Henry VIII Eating a Turkey Leg 59
    Play it Again, Sam 64
    It's a Beautiful Day in the Neighborhood 67
    Shaggy's Adam's Apple 69
    Brittney Spears' Headset.. 71
    Alexander Hamilton Was or Was Not a U.S. President........ 74

CHAPTER 4: Mandela Effects and Misspellings..................... 77
    Kit Kat vs Kit-Kat.. 77
    Febreze vs Febreeze 78
    Jif vs Jiffy 79
    Looney Tunes vs Looney Toons 80

Oscar Mayer vs Oscar Meyer 82
Skechers vs Sketchers 82
Froot Loops vs Fruit Loops 83
Double Stuf Oreos vs Double Stuff Oreos 84
Flintstones vs Flinstones 84

CHAPTER 5: Psychological Explanations for the Mandela Effect 89
    Collective False Memories and Social Influence 90
    Confabulation 92
    Misleading Post-Event Information 94
    Priming 97
    Social Influence and Collective False Memories 99
    Cognitive Dissonance and the Need for Consistency 100
    Memory Reconstruction and Schemas 102

CHAPTER 6: Psychological Explanations for the Mandela Effect 105
    Parallel Universes 106
        Parallel Universes Overview 106
        Parallel Universes and the Mandela Effect 107
    Alternate Realities 109
        Alternate Realities Overview 109
        Alternate Realities and the Mandela Effect 109
    Reality is a Simulation 111
        Reality is a Simulation Overview 111
        Glitches in the Simulation as an Explanation for Mandela Effects 112
    Powerful Entities Controlling Our Perception 113
        Powerful Entities Controlling Our Perception Overview ... 113
        Powerful Entities and Mandela Effects 115

CHAPTER 7: The Internet's Impact on Mandela Effects 119
    Social Media Amplification 120
    Digital Echo Chambers 121

Memes and Viral Content.. 123
Misinformation and Fake News 124
The Role of Influencers and Celebrities 125
Online Debates and Discussions.. 126
Internet Archives and Search Engines 127

Conclusion... 129

Endnotes.. 133

Introduction

We live in an age unprecedented. Technology has made it sometimes impossible to tell reality apart from lies. Deep fakes can make it seem like famous people are doing or saying something they never would. Photo and video editing make spotting aliens and ghosts seem like a daily experience. AI can create fantastical homes, products, and more that when seen by the average viewer can be very, very convincing. Where we now have to question everything, the one thing we should be sure of is our own memory.

Right?

Maybe not.

Enter the Mandela Effect.

I remember my first experience with the Mandela Effect. I was actually looking for books for my two young children—Zack and Brittany—in a local used bookstore. I wanted to share with them the books I loved as a child, especially the Bearenstein Bears books. I remembered those books being some of the first I was able to read all by myself. When I asked the gentleman working there if they had any of this beloved series, he corrected my pronunciation.

"You mean the BearenSTAIN Bears?" he asked.

And that's where I first fell down a Mandela Effect rabbit hole, five years before the term was even coined.

Even with a copy of *The Big Honey Hunt* from the 1960s in my hand and looking at the AIN ending of the amazing author's last name firsthand, I still couldn't believe it. How could I be so sure of something that was so obviously wrong?

That's what we'll explore in the pages that follow.

Despite the diverse interpretations and hypotheses surrounding the Mandela Effect, one thing remains certain: it serves as a potent reminder of the fragility of memory and the fallibility of human perception. As we embark on a journey to unravel the mysteries of the Mandela Effect, it becomes evident this phenomenon transcends mere curiosity, offering profound insights into the workings of the human mind and the nature of reality itself. Through rigorous inquiry, critical analysis, and open-minded exploration, we begin to peel back the layers of illusion and glimpse the elusive truths that lie beneath—unraveling the reality of the Mandela Effect.

Are these events merely the human mind playing tricks on us or are there more nefarious things going on?

CHAPTER 1:
What is the Mandela Effect?

The Mandela Effect is a term coined by paranormal enthusiast Fiona Broome in 2009. It refers to a curious phenomenon where a large group of people remember an event or fact differently from the documented historical record. While the concept gained widespread attention in the digital age, its roots trace back to a specific incident involving the former South African president, Nelson Mandela. Many individuals vividly recall Mandela dying in prison during the 1980s, even though he was released in 1990 and went on to become the president of South Africa. This collective misremembering of Mandela's fate sparked intrigue and debate, laying the foundation for what would later be known as the Mandela Effect.

The Mandela Effect is not limited to political figures like Nelson Mandela; it extends to various cultural, historical, and pop culture references, capturing the imagination of millions worldwide. The origins of the Mandela Effect are shrouded in mystery, with scholars and researchers offering divergent explanations for its occurrence. Some attribute it to the fallibility of human memory, suggesting our recollections are susceptible to distortion and manipulation over time. According to this view, factors such as suggestion, social influence, and cognitive biases can contribute to the formation of false memories, leading individuals to believe in events that never occurred or details that never existed.

Others entertain more speculative theories, delving into the realms of quantum physics, parallel universes, and alternate timelines. Proponents of these hypotheses propose the Mandela Effect may be indicative of reality shifts or glitches in the matrix, where discrepancies between individual memories and objective reality point to the existence of multiple dimensions or parallel worlds. While these theories remain speculative and lack empirical evidence, they underscore the profound implications of the Mandela Effect and its potential to challenge our understanding of reality.

Regardless of its origins or explanations, the Mandela Effect continues to captivate the public's imagination, prompting countless individuals to question their perceptions of the world around them. As we delve deeper into the mysteries of this phenomenon, it becomes increasingly clear the Mandela Effect is more than just a quirky anomaly—it's a window into the complexities of human cognition, the nature of memory, and the elusive truths that lie beyond our grasp.

Cognitive Biases and the Mandela Effect

Understanding cognitive biases is particularly crucial when examining phenomena like the Mandela Effect, where collective misremembering of events or details occurs. One significant aspect of cognitive biases in the context of the Mandela Effect is their role in shaping and reinforcing false memories. Biases such as the misinformation effect, where exposure to misleading information can alter memory, can contribute to the proliferation of incorrect recollections among groups of people. By recognizing these biases, we can better understand why and how false memories emerge and spread, shedding light on the mechanisms behind the Mandela Effect.

Moreover, cognitive biases play a significant role in the interpretation and validation of memories related to the Mandela Effect. Biases such as confirmation bias, where individuals tend to seek out and interpret information in a way that confirms their existing beliefs, can lead people to selectively recall details that align with their misconceptions. This bias can perpetuate the persistence of false memories by reinforcing individuals' confidence in their recollections, even in the face of contradictory evidence. Understanding these biases is essential for critically evaluating the validity of memories associated with the Mandela Effect and discerning between genuine recollections and false impressions.

Furthermore, cognitive biases can influence how individuals respond to discrepancies between their memories and objective reality in the context of the Mandela Effect. Biases such as cognitive dissonance, where people experience discomfort when holding contradictory beliefs or attitudes, can lead individuals to rationalize or dismiss evidence that contradicts their memories. This tendency to prioritize cognitive consistency over accuracy can impede efforts to correct false beliefs and perpetuate the spread of misinformation. Recognizing the role of cognitive biases in shaping responses to conflicting information is crucial for addressing misconceptions associated with the Mandela Effect effectively.

Additionally, understanding cognitive biases can inform strategies for mitigating the impact of false memories and combating misinformation related to the Mandela Effect. By educating the public about the prevalence and mechanisms of cognitive biases, we can empower individuals to critically evaluate their own memories and assess the reliability of information they encounter. Promoting media literacy and critical thinking skills can help inoculate people against the influence of misleading narratives and reduce the susceptibility to false memories. By fostering a greater awareness of cognitive biases, we can cultivate a more discerning and informed society capable of navigating the complexities of memory and perception.

The importance of understanding cognitive biases in the context of the Mandela Effect cannot be overstated. These biases not only contribute to the formation and perpetuation of false memories but also shape individuals' responses to contradictory information and influence the spread of misinformation. By recognizing and addressing these biases, we can gain insights into the underlying mechanisms of the Mandela Effect, develop strategies for evaluating the validity of memories, and promote critical thinking skills to combat misinformation effectively.

CHAPTER 2:
Understanding Memory and Perception

Understanding memory and perception is essential when exploring phenomena like the Mandela Effect, where collective misremembering of events or details occurs. Memory and perception are intricate processes influenced by various factors, including individual experiences, cognitive biases, and external influences. The Mandela Effect highlights the complexities of memory formation and the fallibility of human perception, underscoring the importance of critically examining the reliability of our recollections. By delving into the mechanisms behind memory and perception, we can gain insights into why and how false memories emerge, providing valuable context for understanding the Mandela Effect and its impact on our understanding of reality.

Memory Formation and Retrieval

Memory formation and retrieval are intricate processes that involve various stages and mechanisms. Encoding, the first stage of memory formation, occurs when sensory information is transformed into a form that can be stored in the brain. This process involves the translation of sensory input into neural codes, which are then stored in different areas of the brain depending on the type of memory being formed. For example, visual memories may be stored in the occipital lobe, while auditory memories may be stored in the temporal lobe.

Once encoded, memories are stored in the brain through consolidation, a process that stabilizes and strengthens neural connections associated with the memory. Consolidation occurs primarily during sleep, with memories being transferred from short-term to long-term storage. This process involves the reactivation and strengthening of neural circuits through synaptic changes, making the memory more resistant to forgetting.

Retrieval is the process of accessing stored memories when needed. It involves the reactivation of neural circuits associated with the memory and the reconstruction of the memory from stored information. Retrieval can be influenced by various factors, including the context in which the memory was encoded, emotional state, and the presence of cues or triggers that prompt memory recall.

Overall, memory formation and retrieval are dynamic processes shaped by a combination of biological, psychological, and environmental factors. Understanding these processes is crucial for comprehending how memories are created, stored, and recalled, as well as the potential sources of error and distortion that can impact memory accuracy.

Factors Influencing Memory Accuracy

Memory accuracy can be influenced by a variety of factors, both internal and external, that affect how information is encoded, stored, and retrieved in the brain. One key factor is attention, as focused attention is necessary for effective encoding of information into memory. When individuals are distracted or multitasking, their ability to encode information accurately may be compromised, leading to reduced memory accuracy. Additionally, emotional arousal can impact memory accuracy, with heightened emotional states enhancing the encoding and storage of emotional memories but potentially impairing memory for neutral information.

Another factor impacting memory accuracy is the presence of contextual cues during encoding and retrieval. Memories are often tied to the context in which they were formed, so environmental cues present during encoding can serve as retrieval cues later on. For example, studying for an exam in the same room where the exam will be taken can enhance memory recall due to the context-dependent nature of memory. Similarly, the presence of retrieval cues, such as familiar smells or sounds, can trigger memory recall by activating associated neural networks.

The passage of time also plays a significant role in memory accuracy, as memories can be subject to decay or interference over time. Decay refers to the gradual fading of memories over time if they are not retrieved or rehearsed, while interference occurs when new information disrupts the retrieval of existing memories. Both decay and interference can contribute to memory errors and inaccuracies, particularly for memories that have not been well-consolidated or are not frequently accessed.

Individual differences in cognitive abilities and strategies can also impact memory accuracy. For example, individuals with stronger working memory capacities may be better able to encode and retain information accurately, while those with weaker working memory may experience more difficulties in memory retrieval. Additionally, individuals may employ different mnemonic strategies, such as visual imagery or chunking, to enhance memory encoding and retrieval, with varying degrees of success depending on the individual and the task.

Finally, the suggestibility of memory is a crucial factor to consider when evaluating memory accuracy. Suggestibility refers to the susceptibility of memory to distortion or manipulation by external influences, such as leading questions, misinformation, or social pressure. Memories can be altered or contaminated through suggestive techniques, leading individuals to incorporate false details or misinformation into their recollections. Understanding the factors that impact memory accuracy is essential for recognizing the limitations of memory and mitigating the risk of memory errors and distortions in various contexts.

Role of Perception in Shaping Our Reality

Perception plays a fundamental role in shaping our reality by interpreting and organizing sensory information from the environment into meaningful experiences. It involves the complex process of selecting, organizing, and interpreting sensory input to construct a coherent understanding of the world around us. Perception is not a passive reflection of the external world but rather an active and dynamic process influenced by factors such as expectations, beliefs, and prior experiences.

One key aspect of perception is selective attention, which allows individuals to focus on specific aspects of their environment while filtering out irrelevant or distracting information. Selective attention helps individuals prioritize and process relevant sensory input, enabling efficient perception and decision-making in complex environments. However, selective attention can also lead to perceptual biases and errors when individuals fail to notice important information due to cognitive limitations or biases.

Another important aspect of perception is perceptual organization, which involves the grouping and interpretation of sensory information into meaningful patterns and objects. Gestalt principles, such as proximity, similarity, and closure, guide perceptual organization by grouping elements based on their spatial and structural relationships. Perceptual organization helps individuals make sense of complex visual scenes and facilitates the recognition of objects and patterns in the environment.

Perceptual constancy is another crucial concept in perception, referring to the tendency to perceive objects as stable and consistent despite changes in their sensory input. For example, size constancy allows individuals to perceive an object as maintaining a constant size regardless of its distance from the observer, while shape constancy enables the recognition of objects despite changes in their orientation or perspective. Perceptual constancies contribute to the stability and continuity of our perceptual experiences, allowing us to navigate and interact with the world effectively.

Additionally, perception is influenced by top-down processes, such as expectations, beliefs, and cultural factors, which shape our interpretation of sensory information. These cognitive and contextual factors influence how we perceive and interpret ambiguous or uncertain stimuli, leading to perceptual biases and illusions. Understanding the role of perception in shaping our reality is essential for recognizing the subjectivity and limitations of our perceptual experiences and for fostering empathy and understanding in interpersonal interactions.

In the end, memory and perception are intricate processes that significantly contribute to the phenomenon known as the Mandela Effect. Our memories are not infallible recordings of past events but rather reconstructive processes influenced by various factors such as cognitive biases, social influences, and perceptual mechanisms. Perception plays a crucial role in shaping our interpretation of the world around us, guiding how we select, organize, and interpret sensory information. Together, memory and perception contribute to the formation and propagation of collective false memories, giving rise to intriguing instances of the Mandela Effect. By understanding the complexities of memory formation, retrieval, and perception, we can gain valuable insights into the nature of human cognition and the fascinating ways in which our minds construct reality.

CHAPTER 3:
Famous Mandela Effects

In this chapter, we delve into the intriguing world of the Mandela Effect, exploring notable examples that have captured the imagination and curiosity of many. Let's start to unravel some of the most perplexing instances of the Mandela Effect and examine the fascinating dynamics at play behind these shared false memories.

Nelson Mandela's Death

Let's start this unraveling of reality with the Mandela Effect's namesake—Nelson Mandela.

Nelson Mandela's Death is the group false memory researcher Fiona Broome used to coin the term "Mandela Effect." It is one of the most well-known examples of collective memory distortion and revolves around the widespread belief Nelson Mandela, the anti-apartheid revolutionary, died while imprisoned in the 1980s. In reality, Mandela was released from prison in 1990 and went on to serve as President of South Africa from 1994 to 1999. History reports he passed away in December 2013.

Despite this historical timeline, many people vividly recall hearing about Mandela's death decades earlier, sparking speculation and debate about the nature of memory and perception. Some recall the funeral, even with a eulogy given by his widow, a long funeral procession and visiting dignitaries. Some even remember riots following the anti-apartheid leader's death. Many people recall hearing about his death in school, especially social studies classes. Other report hearing about Mandela's death on the news.

> Why do so many remember a death history reports didn't occur for three more decades?

One explanation for the misremembering of Nelson Mandela Death lies in the phenomenon of confabulation, where individuals unknowingly fabricate false memories to fill gaps in their recollection. In the case of Mandela's death, it's possible people conflated his lengthy imprisonment with his passing, leading to the creation of a false narrative in their minds. Additionally, Mandela's iconic status as a global symbol of resilience and resistance may have contributed to the proliferation of misinformation surrounding his supposed demise. The emotional impact of Mandela's life story, coupled with the dramatic political and social upheavals of the apartheid era, could have made his supposed death a memorable event in the minds of many.

Moreover, the prevalence of media coverage and historical accounts detailing Mandela's struggle against apartheid and subsequent rise to power may have inadvertently reinforced the false memory of his early death. Images of Mandela's imprisonment and the global campaign for his release were seared into the collective consciousness, creating a potent narrative that transcended individual recollections. As a result, people may have internalized the idea of Mandela's death during his years of incarceration, perpetuating the misconception despite contradictory evidence.

Perhaps it was misconstruing news reports about Mandela's health that led people to mistakenly report his death. *The New York Times* reported in 1988 about Mandela's struggle in prison with tuberculosis. At that time, Mandela had been coughing up blood and three quarts of fluid had to be removed from his lungs. This resulted in renewed efforts to get him released from prison.

> The main argument put forward by advocates for his release within the South African Government is that if Mr. Mandela died in prison, it could have catastrophic consequences and that he is worth more to the African National Congress in jail than he would be as a free man[i].

Could news reports like this be the reason why people thought Mandela had died? Remember in the 1980s we didn't have pervasive home computers, 24/7 news stations, cell phones, social media, etc. to get news reports directly. So was it like a childhood game of "telephone" where the message gets skewed with each retelling?

I, personally, don't recall where I heard the Mandela had died in prison specifically, but I was very surprised when I heard about his historical death in 2013. I remember thinking, "Wait… didn't he already die?"

Berenstein Bears vs Bearenstain Bears

As I mentioned, the Berenstain Bears vs. Berenstain Bears Mandela Effect was the first one I personally encountered. At its core, this peculiar memory discrepancy revolves around the spelling of the surname of the beloved bear family featured in the popular children's book and TV series. While the official spelling is "Berenstain," many people (myself included) distinctly recall it being spelled as "Berenstein." This discrepancy has led to widespread debate and speculation, with some attributing it to a simple misspelling while others see it as evidence of parallel universes or alternate realities.

And I'm not alone in this memory. As related by a science blogger named "Reece," thousands of people with the name of Bearenstein recall clearly being teased as children they were somehow related to the famous Bears. Others recall debating why it was pronounced "steen" and not "stine" as pronounced with Frankenstein, Einstein, R.L. Stein, etc.. Still others talk about using the mnemonic of Papa Bear with a stein of beer, to correlate with the last name spelling[ii]. Clearly, I am not alone in remembering the name as Bearenstein.

One possible explanation for why so many people misremember the spelling of the Berenstain Bears' surname is rooted in the nuances of human memory. Our brains are not infallible recording devices; rather, they construct memories based on a combination of sensory input, prior experiences, and contextual cues. In the case of the Berenstain Bears, individuals may have encountered the name at a young age when their cognitive abilities were still developing. This developmental stage, coupled with the phonetic similarity between "Berenstain" and

"Berenstein," could have led to the incorrect encoding of the surname in memory.

Furthermore, the prevalence of other surnames ending in "-stein" may have contributed to the confusion surrounding the spelling of the Berenstain Bears' surname. Names like "Einstein" and "Frankenstein" are well-known and frequently encountered in various cultural contexts, potentially leading individuals to default to the more familiar spelling when recalling the Berenstain Bears' name. Additionally, the passage of time and the dissemination of information through various media channels may have reinforced the incorrect spelling in the minds of individuals who grew up with the books.

Cultural and linguistic influences may also play a role in shaping individuals' memories of the Berenstain Bears' surname. Depending on regional accents or dialects, certain vowel sounds may be pronounced differently, leading individuals to interpret the spelling of words in distinct ways. The pronunciation of "Berenstain" with a long "E" sound may have led some individuals to perceive it as "Berenstein," especially if they were not familiar with the correct pronunciation. Additionally, the association between "-stein" endings and certain cultural or historical figures may have further contributed to the misremembering of the surname.

In fact, in a *National Post* interview, the son of the late Stan and Jan Bearenstain—Michael Bearenstain—commented on how his father had experienced people mistaking his name, even as a child.

> Well, actually the earliest I know about it is in my mother and father's autobiography where my dad wrote a section about when he was in elementary school. His elementary school

teacher said that his name was spelled incorrectly and that she was changing it to "Berenstein," and that she wouldn't recognize the spelling of his name in her class because there was no such name. So it goes back pretty far, the issue. And when I was a kid growing up, nobody pronounced it correctly. I never even tried to get people to pronounce it correctly. They always said "Berensteen" or "Bernstein" or something. I never thought much about it at the time. I just figured that, you know, people pronounce things incorrectly, and that's just the way it is. It's not a new issue, it's just a common phenomenon that happens to people with oddly spelled names[iii].

Could it simply be people, myself included, are so used to seeing last names spelled "ein" and that then being reinforced with the "een" pronunciation, that we all simply didn't pay attention to how the name was actually spelled? Could this expectation have affected our perception?

My last name, Wylie (pronounced Wile E... like the Coyote), is not a commonly spelled version of this pronunciation. I've had people spell it the more common Wiley, when I've just spoken my name—such as at a hostess stand at a restaurant—however, I've never had someone who has seen my name properly spelled in the past say, "Whoa! You spell your name W-Y-L-I-E? I thought I remembered seeing it spelled W-I-L-E-Y?" So is an odd spelling really what's made hundreds of thousands of people swear there's an EIN at the end of the famous Bears' name ?

Or is there something more going on?

Watching reruns of the short-lived (two seasons) *Bearenstain Bears* show, it does sound like they say Bearen-*STEEN*... kind of. Actually, to my ear, it sounds like they're saying Bearen-*SHTAEN*... with a light A sound hidden in there under the E, if that's what I listen for. But, at first listen, it does sound like *STEEN*, especially with Papa's gruff voice[iv]. So could this be why some misremember it? Possibly. Especially those who grew up watching this series. Many likely weren't really paying attention to the title of the show as it came on, especially as children.

But that doesn't explain why I clearly remember the E version versus the A. The TV series came out in 1985. I was 16 and definitely had not watched the show. Fast forward over a decade later in the early 2000s when I went hunting for the easy-to-read bear family book series I hoped to share with my young kids, and without ever having heard of the series, I thought it was Bearenstein. Although the TV show may have influenced some people's memories, it hadn't been a factor in mine.

Even George Takei (insert fangirl swoon here) remembers it as Bearenstein. Check out his Facebook post from 2015[v]:

> Is it just me, or did they used to spell it BerenSTEIN Bears? How is it now spelled BerenSTAIN Bears? Is this a conspiracy?

Right, George? Right?! What is going on?

The popular TV show "The Big Bang Theory" referenced the Berenstain Bears as the "Berenstein Bears" as well. Perhaps this contributed to the widespread Mandela Effect where people remember the name spelled differently. However, although I personally have seen many episodes of this show, I don't recall this reference, so I know in my specific instance of remembering the name as Bearenstein, it wasn't due to this show.

Figure 1: First Bearenstain Bears book published in 1962[vi]

Where the Heck is New Zealand?

The Mandela Effect surrounding the location of New Zealand is a curious phenomenon. While the majority of people correctly identify New Zealand as an island nation situated southeast of Australia, there exists a notable minority who recall it being located in a different geographical position. Instead of its actual location, some individuals distinctly remember New Zealand being located northeast of Australia, closer to Papua New Guinea and the Solomon Islands.

I happen to be one of these people. To be fair, geography is probably one of my poorest areas of knowledge though. So although this one niggled at my brain uncomfortably, as Mandela Effects tend to do, I tried to pass it off as just my incompetence at world geography.

Then I I asked my husband, Grant, where he thought New Zealand is. Now Grant is much more skilled at geography, so I figured he'd get it right. I held up my phone to represent Australia on the invisible map in front of us. I explained up would be north, down would be south, right would be east, and left would be west.

"If this is Australia," I said as I pointed to my phone being held up in mid-air, "where is New Zealand?"

"Right here," he said quickly pointing to the 2 o'clock position from my phone-Australia—northeast… just like I had mistakenly thought! So why was this happening? Because we're not alone. A lot of people thought New Zealand was located here:

Figure 2: New Zealand placement for those of us suffering from this Mandela Effect[vii]

One possible explanation for the Mandela Effect regarding the location of New Zealand lies in cartographic anomalies and cognitive biases. Maps and globes are commonly used tools for representing the world's geography, but they are not infallible and can sometimes distort or misrepresent spatial relationships. The Mercator projection, for example, exaggerates the size of landmasses near the poles, leading to distortions in the perceived distances between regions. Inaccuracies in map representations, coupled with the brain's tendency to prioritize familiar or dominant information, may have contributed to the formation of false memories regarding New Zealand's location.

Furthermore, cultural influences and media portrayals can play a significant role in shaping our perceptions of geographic regions. New Zealand's relatively isolated position in the Pacific Ocean, coupled with its unique cultural identity and stunning natural landscapes, have made it a popular backdrop for films, television shows, and literature. However, these portrayals are not always geographically accurate, and individuals may inadvertently internalize fictionalized depictions of New Zealand's location based on their exposure to media content. The power of suggestion, coupled with the human brain's capacity for associative memory, could have influenced the formation of false memories regarding New Zealand's position on the map.

To complicate matters, many maps simply leave New Zealand off their depiction of the world completely. Let's take a look at a couple...

Figure 3: Missing New Zealand[viii]

Figure 4: IKEA map with missing New Zealand[ix]

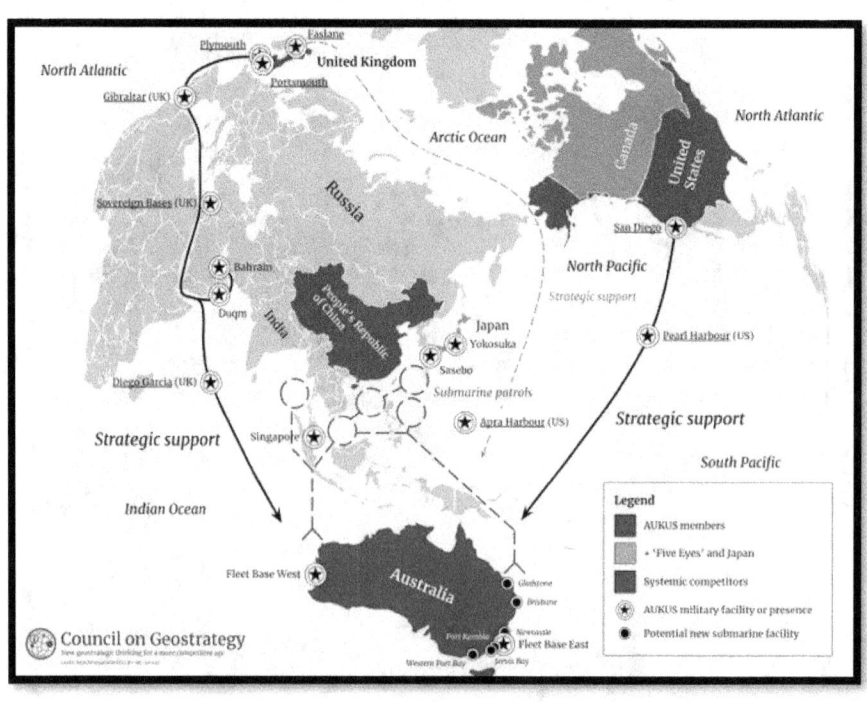

Figure 5: Council on Geostrategy's X post missing New Zealand.[x]

And sometimes New Zealand is placed oddly to the west of Australia, like in the following example.

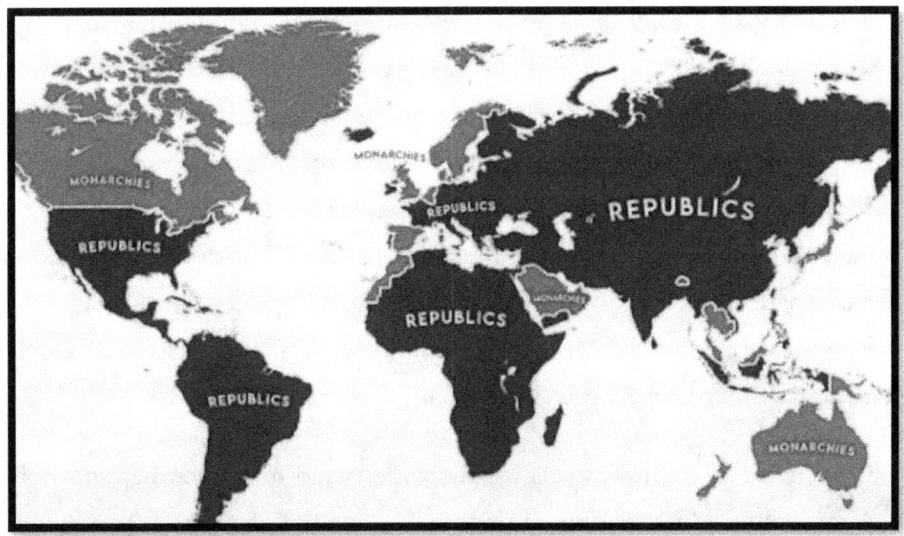

Figure 6: Although this map has New Zealand on it, it's too the west of Australia for some reason.[xi]

Another contributing factor to the Mandela Effect surrounding New Zealand's location may stem from the brain's inherent tendency to fill in gaps in information and construct coherent narratives. When faced with incomplete or ambiguous spatial cues, the brain relies on existing knowledge and contextual clues to construct a mental representation of the world. In the case of New Zealand, individuals who lack firsthand experience or detailed knowledge of its precise location may rely on heuristic reasoning and associative thinking to infer its position relative to other familiar landmarks or regions. This reliance on mental shortcuts and schema-based processing can sometimes lead to inaccuracies or distortions in spatial perception, contributing to the prevalence of false memories regarding New Zealand's location.

Monopoly Man's Missing Monocle

The Mandela Effect surrounding the Monopoly Man and his supposed monocle has puzzled many of the game's enthusiasts. In the iconic board game Monopoly, the character known as Rich Uncle Pennybags (Milburn Pennybags), often referred to as the Monopoly Man or Mr. Monopoly, is depicted as a portly, mustachioed man wearing a top hat, a suit, and a monocle—or so many people believe. However, upon closer examination of the official Monopoly game pieces and artwork, it becomes apparent the Monopoly Man does not actually wear a monocle, contrary to what numerous individuals vividly recall.

One of the possible reasons for the widespread misremembering of the Monopoly Man wearing a monocle can be attributed to cultural stereotypes and media portrayals. The image of a wealthy, upper-class gentleman sporting a monocle has long been ingrained in popular culture as a symbol of sophistication, refinement, and aristocracy. Countless depictions in cartoons, advertisements, and literature have perpetuated this stereotype, leading to a collective association between monocles and affluent characters. Think about fancy Mr. Peanut. Or does anyone else remember the Mayor of Townsville, from the *Powerpuff Girls*? His top hat, fancy suit, and monocle gave him the stereotypical 'rich older guy' look. As a result, it's possible when people think of a wealthy character like the Monopoly Man, the mental image of a monocle often accompanies it, despite the lack of empirical evidence to support this association.

However, the reason I think is most likely the cause of this particular Mandela Effect is…

        The Monopoly Man DOES have a monocle!

At least sometimes.

There have been numerous instances of Mr. Monopoly popping up with a monocle. A 1994 version of Monopoly Junior clearly has Mr. Monopoly on the 2 note with a monocle. A Dutch version of the game also featured the same monocled currency. Interestingly, it was only the 2-note that had the monocle.

Figure 7: A 1994 Monopoly Junior game with a monocled Mr. Monopoly on the 2 note[xi].

Figure 8: A Dutch version of Monopoly Junior with the infamous monocled Mr. Monopoly on the 2 note[xiii].

Figure 9: Only the 2 note featured Mr. Monopoly with his monocle[xiv].

Adaptations of Mr. Monopoly also often include him wearing a monocle which could further cause the skewed memory of the general public. This luxury magazine featured a rendering of Mr. Monopoly on their cover in 2009.

Figure 10: Mr. Monopoly on the cover of a luxury magazine, with a monocle[xi].

During the Equifax hearings in 2017, when then CEO Richard Smith was testifying before the Senate Banking, Housing and Urban Affairs committee, someone dressed as Mr. Monopoly with a monocle photobombed the beleaguered former CEO.

Figure 11: Mr. Monopoly with monocle photobombing Richard Smith during the Equifax hearings[xxvi].

The way Mr. Monopoly has been drawn, over the years, almost looks like he's wearing a monocle, even when he isn't, which may contribute to the memory of him with a monocle. He has little, beady eyes and very rounded, arched eyebrows. With his images often being seen from the side, with more of a focus on one eye than the other, at first glance it's pretty easy to see how someone could mistake his eye as having a monocle. Check out the first depiction of Rich Uncle Milburn Pennybags, as he appeared on the cover of a game dedicated to him—Rich Uncle.

Figure 12: Rich Uncle Pennybags AKA Mr. Monopoly on the cover of his game from 1946[xvii].

Could you see how someone might glance at this image and then remember a monocle that didn't exist?

Of course, Hasbro muddling the memory waters hasn't helped the debate. On their official Monopoly Facebook page, they posted:

I'm not above accessorizing with my mighty monocle.

Along with the following photo:

Figure 13: Facebook post on Hasbro's official Monopoly page[xxviii].

Were they poking fun of the monocle/no monocle debate?

Or are they finally acknowledging the monocle
IS a rare accessory this character DOES indeed wear?

Sinbad as *Shazaam!*

The Mandela Effect surrounding the supposed movie "Shazaam" starring Sinbad is another mystery. Despite overwhelming evidence to the contrary, a significant number of people (myself included) vividly recall a family-friendly film from the 1990s in which the comedian Sinbad played a genie named Shazaam. However, no such movie exists in reality, leading to widespread confusion and speculation about the nature of this apparent false memory.

One possible explanation for the phenomenon is the power of suggestion and cultural reinforcement. In the 1990s, Sinbad was a prominent figure in popular culture, known for his comedic performances and appearances in movies and television shows. Who can forget him as the crazed mailman desperate to get his son a Turboman in "Jingle All the Way?" In addition to his self-titled TV series, "The Sinbad Show," Sinbad appeared in 15 different movies or TV shows in the 1990s.

In 1996, a film titled "Kazaam," starring Shaquille O'Neal as a genie, was indeed released. So there was indeed a genie movie released, with a similar sounding title, in the 90s. Some theorize the combination of Sinbad's persona, the popularity of genie-themed entertainment, and the similarity in titles between "Kazaam" and "Shazaam" contributed to the creation of a false memory among certain individuals. However, Sinbad and Shaq are two very different looking guys.

Sinbad is considered tall, at 6'4"; however, Shaq is truly a giant at 7'1" in height. In the 1990s, Sinbad sported a full head of hair, while Shaq kept his dome shaved. Shaquille's facial hair was typically a thin mustache leading down to a very thin chinstrap, at most, during the 90s. However,

Sinbad wore a bit thicker circle of mustache leading down into a goatee. It's hard to believe so many people, including myself, confused these two very different looking men.

Personally, I really don't think this is what happened in my case.

Figure 14: Shaquille O'Neal in the 1990s during his Orlando Magic career[xix].

Figure 15: Sinbad in the 1990s[∞].

The human brain's susceptibility to misinformation and the tendency to rely on heuristics—mental shortcuts or rules of thumb—when processing information can also play a role in the formation of false memories. Psychologists have long observed memories are not immutable records of past events but rather reconstructive processes influenced by various factors such as emotions, beliefs, and external cues. In the case of the "Shazaam" Mandela Effect, individuals may have encountered references to the nonexistent movie through word-of-mouth, online discussions, or media coverage, leading them to inadvertently incorporate this misinformation into their own memories. While this could be why some people remember a "Shazaam" movie, it doesn't explain everyone. Someone had to misremember it at first to start the Mandela Effect, right?

The Fruit of the Loom Cornucopia

The Fruit of the Loom missing cornucopia is, for me, one of the more disturbing Mandela Effects. While the iconic underwear brand's logo features an apple, a cluster of purple grapes, a cluster of green grapes, some white currants, and leaves some people clearly remember these fruits spilling out of a cornucopia. But Fruit of the Loom insists,

"No, there's never been a cornucopia."

Figure 16: Fruit of the Loom logo versus misremembered logo with cornucopia[xxi].

In fact, the company even devoted a webpage to supporting their claims there has never been a cornucopia. On this page, they outline each of the Fruit of the Loom logos over the years.

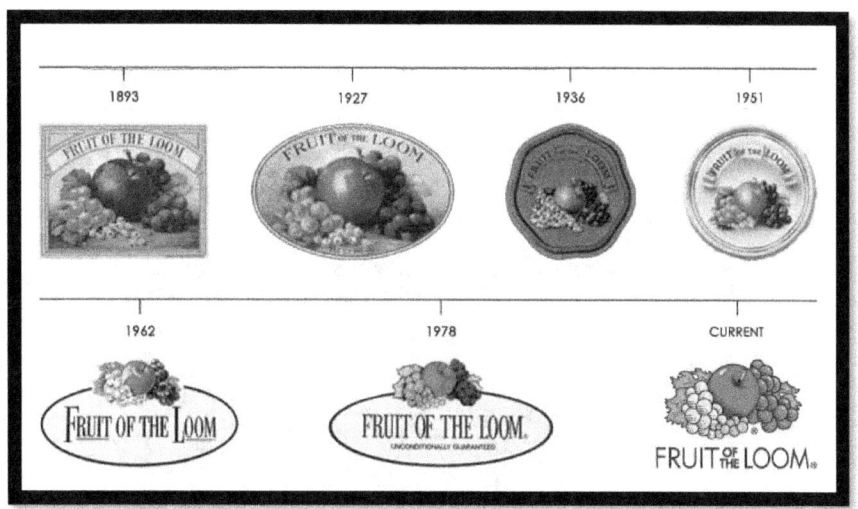

Figure 17: History of Fruit of the Loom logos, per the company[xxii].

Of course, it seems Fruit of the Loom has forgotten about a trademark application they filed in the 1970s. On November 12th, 1973, the Fruit of the Loom Company filed for a trademark for their logo—U.S. Serial Number 73006089. While the drawing submitted did not include a cornucopia, in the Mark Information section, they clearly indicate they were thinking about it. In the Design Search Codes, they indicate four Search Codes:

> 05.09.01 - Berries; Raspberries; Strawberries
>
> 05.09.02 - Grapes
>
> 05.09.05 - Apples
>
> **05.09.14 - Baskets of fruit; Containers of fruit; Cornucopia (horn of plenty)**[xxiii]

Now, why would they have selected the Cornucopia Code, if they hadn't been at least considering placing their fruit in some sort of container?

To add to the mystery, many people have come up with photos of clothing with the cornucopia version of the label.

Figure 18: Photo posted on X of a shirt with Fruit of the Loom label with a cornucopia[xxiv],

Other images have popped up similarly showing instance of the cornucopia logo. However, Snopes, the self-purported Internet debunking site, theorizes all of the images that have surfaced use one of two fake logo images, either as an online photoshop fake or someone who has gone to the trouble of actually making a garment with the fake logo[xxv]. I can concede there are definitely people on the Internet who would love the attention that would come with discovering a garment that would prove there really was a cornucopia in the logo.

But my brain wouldn't leave it alone.

And that's when I came across this…

Figure 19: 1939 photo of blind tiff worker with young child in Washington County, Missouri[xxvi].

45

Now, at first, this photo found on a vintage photography site seems unrelated to our cornucopia dilemma. But then, as user ExploreRealityAlso on Reddit pointed out, one of the newspapers lining the wall has a Fruit of the Loom ad on it[xxvii].

Let's take a closer look to the section of the photo in question...

Figure 20: Close up of Rothstein photo

You can see the Fruit of the Loom logo just under the word "FROCKS." Although the zoomed in section isn't crystal clear, it does kind of look like a cornucopia behind the fruit. Of course, it's hard to say for certain, given the quality of the image, but it could be there.

Sleeping Beauty

I will admit, I'm a Disneyphile. Yes, I'm that adult who goes to the Parks multiple times a year with no children. I eagerly await the release of the next Disney film. And when I can't find anything new to watch on TV, you'll find me watching a Disney movie I've likely seen multiple times before on Disney+. So the Mandela Effect surrounding the quote from *Snow White* as the Evil Queen gazes into her mirror is particularly fascinating to me. Despite being a staple of popular culture for decades, the actual line from the 1937 Disney animated film *Snow White and the Seven Dwarfs* is one of the most controversial Mandela Effects.

Figure 21: The Evil Queen approaching her Magic Mirror, just before she recites the line in question[xxviii].

For many, we remember this infamous line as, "Mirror, Mirror on the wall; who's the fairest of them all." However, in (at least this) reality, the

47

quote is "Magic Mirror on the wall; who's the fairest of them all." It seems a minor differentiation, but for those who remember "Mirror, mirror…" it can feel pretty major.

> So why would someone misremember a line from such a classic movie anyway?

One possible explanation for the misremembered quote is the phenomenon of cultural saturation. "Mirror, Mirror on the wall" has been widely referenced and parodied in numerous adaptations, retellings, and media portrayals of the Snow White story over the years. As a result, the altered version of the line may have become deeply ingrained in popular consciousness, leading people to believe it is the authentic quote from the film. This cultural saturation could have primed individuals to recall the misquoted version of the line, despite its deviation from the original script.

> Of course, we still have to go back and wonder why it was misquoted originally.

Furthermore, the power of collective memory and shared cultural experiences may contribute to the perpetuation of the misremembered quote. As individuals discuss and reminisce about their favorite movies, scenes, and quotes, the shared reinforcement of the incorrect version of the line can solidify its status as the "correct" memory. This phenomenon is compounded by the fact many people first encounter Snow White and the Seven Dwarfs in childhood, during a formative period of cognitive development when memories are particularly susceptible to distortion and influence.

> Perhaps it is just our child brains at fault.

C-3PO's Leg

At this point, it likely won't come as a surprise one of my favorite movie franchises is the *Star Wars* franchise. Growing up, I wanted an ewok for a pet, I had a crush on Han Solo, and I tried (and failed) to put my hair into Princess Leia buns. I do not know exactly how many times I've seen the original three movies, *Star Wars: A New Hope, Return of the Jedi,* and *Empire Strikes Back,* but dozens of times each is a conservative estimate.

How did I miss C-3PO's silver leg?!?

I know I'm not the only one who's asked themselves this question, because this Mandela Effect is pretty hotly debated among pop culture enthusiasts. Many of us, myself included, clearly remember him being completely gold-plated. Yet, sometime between the "remastering" of the older movies and the new-generation movies somehow C-3PO's right lower part of his leg turned silver. And, according to LucasFilm, it's always been that way.

One possible explanation for the misremembering of C-3PO's leg is the selective attention theory. In psychology, selective attention refers to the tendency of individuals to focus on certain aspects of their environment while ignoring others. In the case of C-3PO, fans may have been so captivated by the etiquette and protocol droid's overall appearance, or the storyline of the films, they failed to notice or remember the subtle difference in coloration between his legs.

It has also been theorized the silver leg could be reflecting the color in many shots of its gold counterpart. This could make it look gold. While on Tatooine, the goldish-red dirt of the desert planet could also be reflecting in the silver, masking it to viewers. Also, in many scenes we don't see C-3PO's full legs. The shot is either of him from the waist up or he has his ever-present friend, R2-D2, alongside him blocking his legs from the camera, at least partially.

However, the reality is there are so many depictions of C-3PO with two gold legs, even shots from the film, it's hard to believe his right lower leg has always been silver.

Let's take a look at some of these.

Movie posters show him with two gold legs, fan posters clearly show both legs are gold, and even a documentary entitled the *Making of Star Wars* discusses when C-3PO had his footprints set at the famous Grauman's Chinese Theater and both of his legs are gold.

Even the beloved Star Wars figurines from the 1970s have two gold legs on C-3PO's action figure. If he had one silver leg in the movie, why wouldn't Kenner have incorporated this detail into their licensed toys? Nearly 30 years later, Hasbro also sold C-3PO action figures with two gold legs. So is this really just a false memory for so many, or did LucasFilm add this silver detail in their remastering and newer movies? And if so, why wouldn't they simply admit it?

Figure 22: C-3PO with his right leg looking notably gold[xxix]*.*

Figure 23: Star Wars: A New Hope movie poster with C-3PO and two gold legs[xxx].

Figure 24: C-3PO and other castmates behind the scene, with two gold legs[xxxi].

Figure 25: Close up showing two gold legs.

53

Figure 26: Poster from 1977 official monthly poster magazine showing C-3PO and two very gold legs[xxxi].

Figure 27: C-3PO setting his footprints at Grauman's Chinese Theater in the 1977 documentary about the making of Star Wars with two gold legs[xxxiii].

Figure 28: 1977 C-3PO Large Size Action Figure by Kenner with two good legs[xxxiv].

Figure 29: Hasbro C-3PO action figure from 2006 with two gold legsccxv.

With so many examples of C-3PO with two gold legs, in both physical figurines and photographic evidence, I really don't think this is a true Mandela Effect. I just don't know what the motivation behind saying he's always had a silver leg is. That's where the mystery is!

Dear LucasFilm,
What's the real story with C-3PO's leg?
Sincerely,
Fans Everywhere

"Luke, I am Your Father."

Staying in the Star Wars universe, let's fast-forward a few years to *Star Wars: Episode V – The Empire Strikes Back*. This movie is one of my favorites. I remember the first time I watched it. It actually features one of the first plot twists I remember acknowledging. The twist when it's revealed Darth Vader is Luke's father.... Mind blown!

And that's where this Mandela Effect comes into play.

For many of us, we remember this iconic scene so well.

Darth Vader and Luke are dueling. Darth Vader has the upper hand and slices a battered Luke's hand off with his lightsaber. Vader then tries to entice Luke to join him, telling him he will complete his training if he joins the Dark Side. He then brings up that fated conversation:

"Obi Wan never told you what happened to your father," Vader taunts.

"He told me enough," Luke says as he moves further away. "He told me you killed him."

As the music reaches a crescendo Vader delivers a classic line that has been immortalized in so many pop culture references…

"Luke, I am your father."

But wait! That's NOT the line Dath Vader says! Nope. Despite what you (and I) may remember, the line is actually…

"No. I am your father."[xxxvi]

Definitely NOT "Luke, I am your father."

Figure 30: Darth Vader reaching out to Luke.

So why the discrepancy?

One possible reason for the misremembering of this quote is the way it has been referenced and parodied in popular culture over the years.

Countless films, TV shows, commercials, and memes have referenced the line, often altering it slightly for comedic effect. As a result, the misquoted version, "Luke, I am your father," has become deeply ingrained in the collective consciousness, overshadowing the original line spoken in the film.

Another factor contributing to the Mandela Effect surrounding this quote is the selective attention and reinforcement of false memories. In the context of the film, the revelation of Darth Vader's true identity is a pivotal moment in the narrative, making it likely viewers would focus more on Luke's reaction to the revelation rather than the specific wording of Vader's line. And just a few lines before the misquoted line, Darth does say, "Luke, you do not yet realize your importance." Could our memories be blending the conversation?

The Mandela Effect surrounding this quote highlights the fallibility of human memory, particularly when it comes to recalling details from popular culture. Memories are not static recordings of past events but are instead reconstructive in nature, influenced by various factors such as perception, interpretation, and external cues. The misremembering of the quote "Luke, I am your father" serves as a reminder of the complexities of memory and the ease with which false memories can be formed and perpetuated.

King Henry VIII Eating a Turkey Leg

When I personally think of eating a turkey leg, I think of two things—Renaissance Faires and Magic Kingdom. However, if you asked me, "Is there a painting of Henry the VIII eating a turkey leg?" I would've thought about it for a moment and then said, "Sure!"

And a lot of other people would agree with me.

But I (and all those others) would've been wrong.

Many people vividly recall seeing a painting depicting King Henry VIII of England holding a turkey leg, a symbol of his opulence and gluttony. However, no such painting exists in historical records, and there is no evidence to suggest Henry VIII ever posed for such a portrait. Despite this lack of tangible evidence, the false memory of the painting persists in the collective consciousness, prompting speculation and intrigue among those who believe they have seen it.

One possible reason for the misremembering of this painting is the association of turkey legs with medieval feasting and royalty in popular culture. Turkey legs have long been depicted as a staple of extravagant banquets and royal courts in historical dramas, literature, and folklore. As a result, the image of Henry VIII, a larger-than-life monarch known for his lavish lifestyle and love of food, holding a turkey leg may have become ingrained in people's minds through cultural osmosis, despite its lack of historical accuracy.

In addition to this, I think the most memory damage was done by a 1933 movie entitled *The Private Life of Henry VIII*. In this movie, actor Charles Laughton played the infamous king, portraying him as a gluttonous eater, including messily gobbling down chicken. Although many people today likely haven't seen this actual film, it could have influenced 20th century portrayals on Henry VIII which have endured into the common turkey leg theme we see.

Figure 31: Charles Laughton as King Henry VIII in the 1933 film - The Private Life of Henry VIII[xxxvii].

Just look at the cover of the 2007 book *Terrifying Tudors: Horrible Histories*. It features an illustration of Henry VIII waving a big turkey leg around and shouting, "Heads will roll!" This illustration by Martin Brown is just one example of many that feature the King wielding a turkey (or chicken) leg.

Figure 32: Just one of many of portrayals of Henry VIII holding a turkey leg[xxxviii].

The misremembering of the painting may also be influenced by the nature of actual historical portraits of King Henry VIII—specifically one by Hans Holbein the Younger. This painting was created in 1537 and one of the most iconic. In it, Henry VIII is holding a pair of gloves in his right hand. It's possible people have seen this painting, and then with the combination of other media (TV, movies, parodies, etc.) portraying him with a turkey leg, our minds have just melded the two together.

Figure 33: Portrait of Henry VIII with gloves in hand[xxxix].

In the end, this tendency to conflate historical fact with cultural stereotypes and tropes likely contributed to the formation and perpetuation of false memories over time.

Play it Again, Sam

Humphrey Bogart and Lauren Bacall as Rick and Ilsa in *Casablanca* took a phenomenal screenplay to another level. Their chemistry shone on the silver screen. And, to this day, this 1942 film is a favorite for many. The line, "Play it again, Sam." is one of the most iconic pieces of dialogue to come out of Hollywood.

But it was never uttered.

Crazy, right? I mean there are even companies that have used a play on that line as their company name—Play it Again Sports comes immediately to mind. So what exactly is said?

The scene finds Rick nursing his heartbreak with a bottle of booze when Sam finds him and tries to get him to go home or go anywhere else but drowning his sorrows in his bar.

Rick says, "Of all the gin joints in all the world, she walks into mine," as Sam sits at the piano and begins to play softly, realizing his friend isn't going anywhere anytime soon. Rick asks him what he's playing, and Sam tells him it's something he came up with.

Rick tells him, "Well, stop it. You know what I want to hear."

"No, I don't." Sam quickly replies, trying to avoid playing the song he knows is only going to send Rick further into his depression.

"You played it for her; you can play it for me," Rick insists.

"Well, I don't think I can remember…" Sam starts when Rick quickly and forcibly interrupts him.

"If she can stand it, I can! Play it!"[xl]

Figure 34: Rick listening to Sam playing As Time Goes By[xli].

And then Sam begins to play *As Time Goes By*.

That's it.

So why do so many get it wrong?

One reason for the misremembering of this quote is the way it has been referenced and parodied in popular culture over the years. Countless films, TV shows, commercials, and memes have used variations of the misquoted line, often altering it slightly for comedic effect. As a result, the misquoted version, "Play it again, Sam," has become deeply ingrained in the collective consciousness, overshadowing the original line spoken in the film.

Additionally, the Mandela Effect surrounding this quote may be attributed to the selective attention and reinforcement of false memories. In the context of the film, the line "Play it again, Sam" captures the essence of Rick's request for the pianist to play the song *As Time Goes By*, even though the actual wording differs from what many people remember. Combine this with Ilsa's earlier line where she tells Sam to "Play it, Sam." as she asks him to play *As Time Goes By*, and it becomes easier to see how maybe folks have only seen the movie once (and often times years, if not decades, earlier) could believe the misquote is correct.

The misremembering of the quote may also have been affected by the 1972 Woody Allen film, *Play it Again, Sam*. Allen plays a neurotic, recently-divorced film critic who is obsessed with the movie, *Casablanca*. He tries to emulate Humphrey Bogart as he tries (and fails) to flirt with women. The title of this film definitely didn't help common memory to correctly remember the *Casablanca's* original line. And it makes me wonder if Woody Allen was misremembering the line when he wrote the screenplay, or was it purposefully misquoted and unintentionally spurred a Mandela effect.

It's a Beautiful Day in the Neighborhood

I grew up on PBS, which meant *Sesame Street* followed by *Mister Rogers' Neighborhood*. When my kids were born, they too watched *Mister Rogers'*. For this reason, I've heard the iconic opening song to this show hundreds, if not thousands of times. I can picture the shot of the flyover of the neighborhood until it finally comes to Mister Rogers' house, then the flashing traffic light, and finally Mister Rogers entering himself as he began to sing:

> It's a beautiful day in the neighborhood,
>
> A beautiful day for a neighbor.
>
> Would you be mine?
>
> Could you be mine?
>
> Won't you be my neighbor?

Except, those lyrics aren't quite right.

The first line of the song is actually:

> It's a beautiful day in THIS neighborhood.

This seems like a minor discrepancy, but if you've been singing and/or listening to this song for decades, like I have, how could we not have heard this?

I thought, perhaps, Mister Rogers simply didn't annunciate his opening line well enough and the S sound in "this" wasn't coming through strongly. Mishearing lyrics isn't uncommon. Heck, my mother thought the Creedence Clearwater song was "There's a bathroom on the right." instead of "There's a bad moon on the rise."

So I spent some time listening to the intro song from episodes from the very beginning, in 1968[xlii]. The one thing that astounded me, from the very beginning, Mister Rogers clearly says THIS. It's not muffled. It's not mispronounced. It's clear as day.

So why the confusion?

One reason for the misremembering of this line is the way it has been referenced and parodied in popular culture over the years. Countless films, TV shows, commercials, and memes have used variations of the misquoted line, often altering it slightly for comedic effect. As a result, the misquoted version, "It's a beautiful day in the neighborhood," could have become deeply ingrained in the collective consciousness, overshadowing the original line from the song.

Furthermore, the misremembering of the line may also be influenced by the nature of nostalgic childhood memories and their tendency to be idealized or simplified over time. *Mister Rogers' Neighborhood* holds a special place in the hearts of many people, and the misquoted line effectively captures the sentiment of the show's message about kindness, empathy, and community. Since most of us haven't watched the show since we were children, it's easy to misremember a word or two, especially if it doesn't change the meaning of the lyric. Couple this with reinforcement from others who have also simplified it and misremembered it, and it becomes easier to see how it could happen, without any nefarious or mystical reasoning behind it.

Shaggy's Adam's Apple

Another staple in my household, when I was growing up, was Saturday morning cartoons, which always included *Scooby Doo, Where Are You?* Although all of the gang were key in solving each week's mystery, let's be honest, Scooby and Shaggy were the true stars. Who doesn't remember, shaggy-haired Shaggy, with his lanky arms and legs, stubbly, chin hairs, and his prominent Adam's apple?

Except, he didn't have an Adam's apple!

Figure 35: Shaggy Rogers in the third episode of Scooby Doo, Where Are You?[xliii]

One reason for the misremembering of Shaggy's Adam's apple may stem from his physical appearance and mannerisms. As we talked about Shaggy is often depicted as a tall, lanky, and somewhat awkward teenager with a distinctive voice and exaggerated facial expressions. These characteristics may lead viewers to associate him with typical teenage traits, such as the presence of an Adam's apple, even if it is not explicitly shown in the animation.

But I think the real reason many of us remember Shaggy with an Adam's apple is because of this...

Figure 36: Shaggy meeting the werewolf for the first time[xliv].

When Shaggy gets frightened, he often 'gulps.' This gulping does show the movement of an Adam's apple, indicating his fear. So although his Adam's apple isn't normally prominent, and we only see is super-skinny neck, there are indeed times when his Adam's apple comes out.

Mystery solved!

Brittney Spears' Headset

The Mandela Effect surrounding Britney Spears' "Oops!... I Did It Again" music video is another collective false memory in popular culture I personally experience. Many people, myself included, distinctly remember Britney wearing a headset microphone in the video along with her iconic red latex jumpsuit, on Mars. However, upon reviewing the music video, it becomes clear she does not wear a headset microphone at any point[xlv]. This discrepancy between memory and reality has puzzled fans and sparked widespread discussion about why such a vivid and specific false memory exists.

One possible explanation for this misremembering is the association of headset microphones with Britney Spears' live performances and other music videos. During her concerts and many televised performances, Britney often wore a headset microphone, which became a hallmark of her stage persona. She's also worn other notable red bodysuits on stage, with a headset. This consistent visual element in her live shows might have influenced fans to project the same image onto the "Oops!... I Did It Again" music video, especially during the dance segments that have a stage performance feeling, blending memories of different performances into a single, cohesive image in their minds. These bold visuals of the music video, combined with the high-energy choreography, might lead viewers to unconsciously add the familiar element of a headset microphone, as it seems to fit seamlessly within the context of a high-tech, performance-oriented video. This phenomenon is a testament to how the brain fills in gaps in memory with plausible details, creating a coherent narrative even if it deviates from reality.

Figure 37: Britney Spears wearing red bodysuits during stage performances, reminiscent of the bodysuit she wore in her video, while wearing a headset microphone.[xlvi xlvii]

Furthermore, the power of suggestion and collective reinforcement plays a significant role in perpetuating this false memory. Once a few individuals begin to assert they remember seeing the headset microphone, others may start to question their own memories and conform to the shared recollection. Social interactions, discussions on fan forums, and references in media can amplify this effect, leading to a widespread but inaccurate belief. The Mandela Effect, in this case, highlights the malleability of human memory and the influence of social dynamics on our perceptions of reality.

To complicate matters, there was actual merchandise released for "Oops!... I Did it Again" that included a microphone headset. There were a couple of licensed dolls of Britney, from the video, that included microphone headsets with the doll. Also, many versions of costumes (both licensed and unlicensed) have been released over the past 20-plus years that include a microphone headset. These products have helped reinforce the false memory a headset was present during the video.

Figure 38: Both dolls and Halloween costumes have been released over the years with a microphone headset that wasn't present in the video.[xlviii, xlix, l]

The misremembering of the headset microphone can be attributed to the blending of memories from Spears' live performances, the fitting nature of the imagined detail within the video's context, the reinforcing power of social suggestion, and the power of merchandising with incorrect components. This phenomenon underscores the intricate ways

73

in which our brains construct and reconstruct memories, often leading us to confidently recall details that never actually existed.

Alexander Hamilton Was or Was Not a U.S. President

The Mandela Effect surrounding Alexander Hamilton being remembered as a U.S. President is a fascinating example of how collective false memories can alter historical understanding. Despite never having held the office of President, many people vividly recall Hamilton being one of the early Presidents of the United States. This misconception is particularly intriguing given Hamilton's significant role in the founding of the nation, which may have contributed to the widespread, yet incorrect, belief tat he served as President.

One possible reason for this misremembering is Alexander Hamilton's prominent role in American history. As one of the Founding Fathers, the first Secretary of the Treasury, and the principal author of the Federalist Papers, Hamilton's contributions were foundational to the establishment and development of the United States. His influence on the early financial and political systems may lead people to mistakenly elevate his status to that of a President, conflating his considerable impact with the highest office in the country.

Starting in 1928, Alexander Hamilton appeared on the $10 bill. With other U.S. Presidents, like George Washington, Abraham Lincoln, Andrew Jackson, Thomas Jefferson, and others, on U.S. currency, people may make the assumption all people on U.S. currency are former U.S.

Presidents. This, of course, would be incorrect, as there are others besides Hamilton who have found their likenesses on U.S. money. These include Benjamin Franklin and Sacagawea. However, with a majority of money featuring Presidents, it's very possible people simply aren't aware of the non-U.S. President currency.

Additionally, the blending of Hamilton's achievements with those of actual Presidents from the same era, like George Washington and Thomas Jefferson, might contribute to this Mandela Effect. Both Washington and Jefferson were contemporaries and collaborators with Hamilton, and all three are frequently discussed in the context of the Revolutionary War and the early republic. Given their intertwined narratives, it is easy to see how people might confuse their respective roles, leading to the mistaken belief Hamilton also served as President.

The popularity of the Broadway musical "Hamilton," which dramatizes Alexander Hamilton's life and work, might also play a role in this collective misremembering. While the musical does not explicitly depict Hamilton as a President, its portrayal of his extraordinary contributions to the nation's founding. This could amplify the impression of his prominence, inadvertently reinforcing the false memory. Although, I would think the musical's success would have reminded a broader audience Hamilton was never President, the reality is even with its popularity, a large percentage of people haven't seen the musical. And, likely, people who have seen it have at least a remote interest in U.S. history, and may already correctly remember Hamilton as a Founding Father and not a U.S. President.

CHAPTER 4:

Mandela Effects and Misspellings

Beyond the Bearenstain/Bearenstein Bears debacle, the number of claimed Mandela Effects and misspellings is pretty amazing. Let's take a look at the most common ones people feel aren't what they used to be.

Kit Kat vs Kit-Kat

In 1935, a delicious chocolatey treat was created in London—Rowntree's Chocolate Crisp. Two years later, it made the trip across the pond to the United States and was rebranded Kit Kat. In 1970, Hershey took over production of these yummy snacks[li]. But for some reason, people seem to remember the name of these wonderful crispy, chocolate bars as being Kit-Kat… with a hyphen.

Despite some photoshopped images online, it simply isn't true. Looking at registered trademark information, Kit Kat has always not included a hyphen. However, it does appear that recent iterations have the logo smooshing together the two words to appear more like KitKat. Of course, the official Hershey website refers to the product name all capitalized and with a space—KIT KAT.

Perhaps this is part of the confusion. Referring to brand with several different formats can be confusing. Add to this that the sound of Kit Kat seems to be similar to other commonly hyphenated words such as: tick-tock, ding-dong, see-saw, etc. It's not surprising the general public may have misremembered a hyphen.

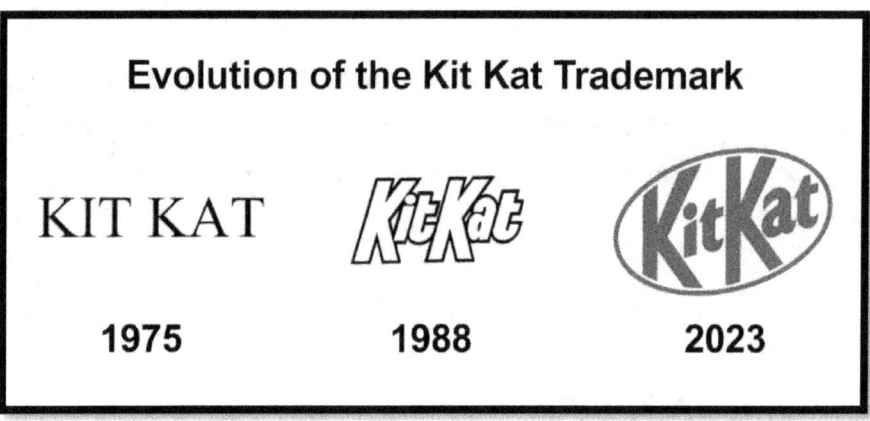

Figure 39: The evolution of the Kit Kat trademarks[lii] [liii] [liv].

Febreze vs Febreeze

Many people distinctly remember seeing the popular air freshener product spelled with a double "e." One possible explanation for this misremembering is the phonetic similarity between "Febreze" and "breeze," which could have subconsciously influenced people to perceive the spelling as "Febreeze." The brain has a tendency to fill in gaps or make connections based on familiar patterns might have contributed to the formation of this false memory, as "Febreeze" appears visually similar to other words ending with "-eeze." Despite the widespread belief in the alternative spelling, the official branding and historical records confirm the correct spelling has always been "Febreze."

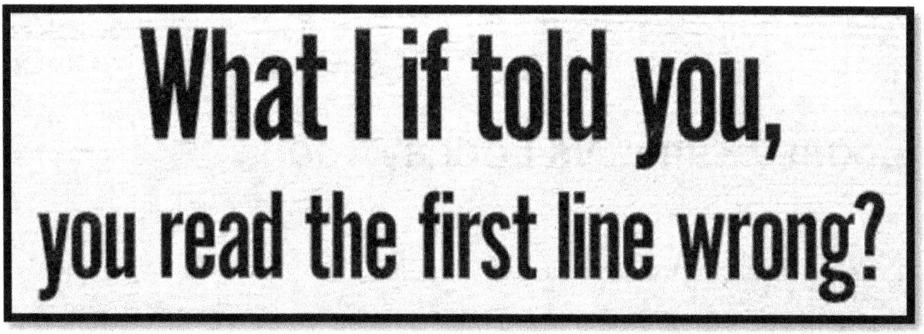

What I if told you, you read the first line wrong?

Figure 40: The brain is always using shortcuts and making connections based on familiar patterns.

Jif vs Jiffy

The common misconception that Jif peanut butter used to be spelled as "Jiffy" is another misspelling Mandela Effect in consumer products. Many individuals vividly remember seeing the product labeled as "Jiffy" on grocery store shelves, leading to widespread confusion and debate. The most likely explanation for this misremembering may be from competitor Skippy.

Jif + Skippy = Jiffy?

I mean it would be the perfect couple name for the two peanut butters, right?

Additionally, the association with the phrase "in a jiffy," meaning quickly or immediately, could have further reinforced the false memory of the brand name being "Jiffy." Despite the widespread belief in the alternative spelling, historical records and official branding confirm the correct spelling has always been "Jif."

Looney Tunes vs Looney Toons

Let's go back to those Saturday morning cartoons. Who didn't watch Looney Tunes? Bugs Bunny, Daffy Duck, Yosemite Sam, and their friends were definitely staples in my home. But, as I kid, I never really paid attention to how to spell Looney Tunes. However, if you had asked me to spell it (before I started researching this book), I likely would've spelled the second word "Toons" simply because that's what they were—looney toons. Wacky cartoons.

Why they are "tunes" and not "toons"… who knows?

Figure 41: Although it would seem to make sense that the Saturday morning favorite cartoons would be called Looney Toons, they are indeed Looney Tunes[iv].

Oh, wait! I know why it's "Tunes" and not "Toons!"

The answer is simple—Disney. Disney Silly Symphonies had become very popular in 1930. This animated series of shorts highlighted musical features. Warner Brothers developed Looney Tunes as a direct competitor to Silly Symphonies. Like Disney's counterpart, Looney Tunes were animated shorts, often shown before feature films, that could utilize Warner Brothers' music library. If you watched their first short, *Sinkin' in the Bathtub*, and substituted Mickey Mouse and Minnie Mouse for Bosko and his girlfriend, it could've easily been a Disney work.

Oscar Mayer vs Oscar Meyer

The widespread misconception that the popular lunch meat brand is spelled "Oscar Meyer" instead of its actual spelling, "Oscar Mayer," is a misspelling Mandela Effect. Many people distinctly remember seeing the brand labeled as "Oscar Meyer" on grocery store shelves. One possible explanation for this misremembering is the pronunciation of the brand's name, with "Mayer" being pronounced MY-er as opposed to MAY-er when spoken aloud. The brain's tendency to rely on phonetic cues and associations could have contributed to the formation of this false memory, as "Meyer" (when pronounced with the long 'I' sound) is a more commonly pronounced surname than "Mayer" with the long 'I' sound in English-speaking countries.

Skechers vs Sketchers

Some of these misspelled Mandela Effects I can understand. I've probably called Jif "Jiffy" more than once, but I can see why I may have accidentally got that one wrong. However, when I look at one popular sneaker brand's correct spelling of its name, it just looks wrong!

Skechers

Now, I've never worn these shoes, nor purchased them for any of my family, but I've seen them in stores. I've also seen their commercials. And for some reason, like many others, I really thought there was a T in the name. Of course, this may simply be because of the word "sketch" and the T in that. Like Mayer versus Meyer, the phonetic spelling is often a possible underlying cause for a misspelled Mandela Effect. Or perhaps it's some alternate universe I've slipped into.

Froot Loops vs Fruit Loops

Going back to the Saturday mornings of my childhood, in addition to watching Scooby-Doo and Looney Tunes, it wouldn't be uncommon to find child me eating a bowl of Fruit Loops...

...or is it Froot Loops?

Despite my memory thinking it is Fruit...like fruit...Loops, the popular cereal since the 1960s is actually spelled Froot Loops. Although there were rumors the cereal changed their name due to legal concerns, since their cereal doesn't actually contain any fruit, historical records show it's always been spelled Froot.

Figure 42: Froot Loops box from 1967[iv].

The reason we might be misremembering this is the simple fact most people know "fruit" is spelled F-R-U-I-T. When we read, we often gloss over words and let our brains make the judgment of what's there, without actually reading. Just like the Febreze example, if our brains expect something, then that's likely what it will initially see and then remember.

Even if it's wrong.

Double Stuf Oreos vs Double Stuff Oreos

Believe it or not, there's no second "F" in Double Stuf Oreos. Crazy, right?

Like Frebreze and Froot Loops, I think this one can also be explained with our brains seeing what we expect to see. Obviously, the correct spelling of "stuff" is with two Fs. In fact, another common misremembering of this product name is Double Stuffed Oreos. This also makes sense, as it would be more grammatically correct to call them "double stuffed."

Flintstones vs Flinstones

Yabba dabba doo! This is one Mandela Effect misspelling I personally did not experience. Although, for me, it has always been *The Flintstones*, with a T, many people swear it used to be spelled "Flinstones"—no T.

In 1959, a 90-second pilot of *The Flagstones,* was created. Creator Joe Barbera discovered there was already a comic series by this name, so changed the show name to *The Flintstones*[lvii]. Given the original name, where flagstone is a type of rock, and the name that stuck, where flint is also a type of rock, it makes sense this show was always known (post Flagstone pilot) by Flintstones. I mean, this is a show about a caveman family in the Stone Ages… so I'm not sure why some swear it was the Flinstones.

The only logical explanation I can think of is if people were pronouncing it "Flinstones" because it is a little more difficult to fully annunciate that first T at the end of "flint" followed by the ST in "stones."

> Go ahead. Give it a try. FlinT-Stones… it definitely doesn't roll off the tongue.

So if people were not pronouncing the first T, that possibly could be the reason why they were misspelling the name in their minds. That's the only possible logical reason I could think of.

At least that's what I thought until I did a quick Trademark search and found this…

85

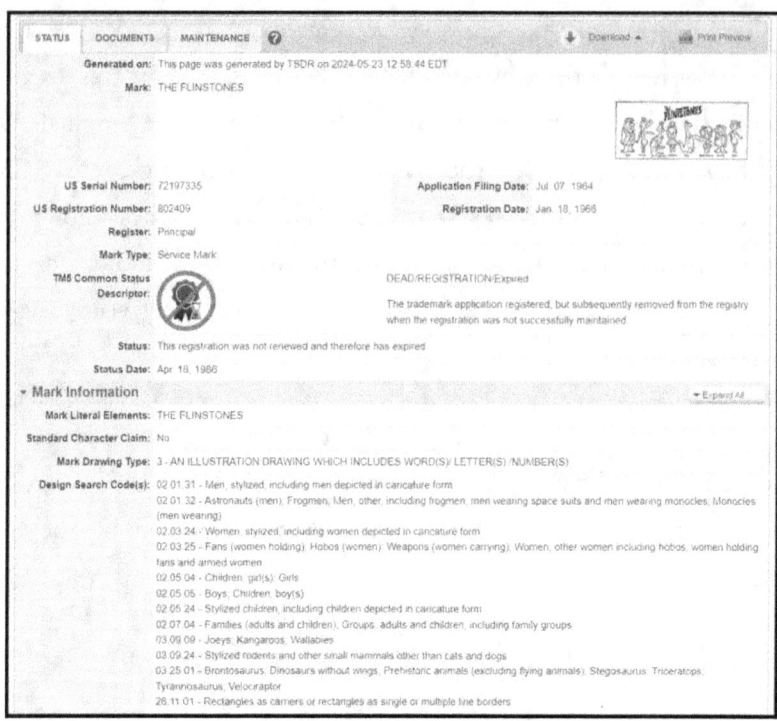

Figure 43: Trademark registrations by Screen Gems Inc. from 1964 to 1966 for The Flinstones[lvii].

Figure 44: Close up of the Flinstones Trademark registration by Screen Gems.

From 1964 to 1966, Screen Gems, Inc. held a registered trademark for THE FLINSTONES—no T. From 1957 through 1966, Screen Gems, Inc. was one of the stakeholders of Hanna-Barbera (holding 20% of the company) and acted as their distributor for their productions[lix]. Although they had filed the T-less version of the name as a trademark, the attached drawing of The Flintstones did feature a T in their name. So I'm not sure why they had a trademark without the T. And I could find no other (not photoshopped) instances of the name on cartoons or merchandise without the T.

So the mystery continues.

CHAPTER 5:
Psychological Explanations for the Mandela Effect

At this point, you've read about many Mandela Effects people have reported and, chances are, there are a few (if not a lot) you read and thought, "No, I don't remember that differently at all." But I'm betting there were some that made you say, "Yes! It definitely used to be …!" Or maybe you thought, "Wait a minute! It's NOT ….?!?!" If you're like me, some of these are so engrained in your memory, when you hear something to the contrary it just doesn't sit right.

So why is this happening to us? There are quite a few theories. Let's start by exploring the psychological explanations that might be behind these sometimes-unsettling Mandela Effects.

Collective False Memories and Social Influence

The phenomenon of collective false memories, often exemplified by the Mandela Effect, is a powerful illustration of the psychological impact of social influence on memory. When large groups of people share the same incorrect memory, it underscores how social interactions and cultural factors can shape and sometimes distort our recollections. Social influence plays a crucial role in the formation of collective false memories, as people are highly susceptible to the suggestions and recollections of others. This can occur through various means, such as conversations, media consumption, and even social media, where misinformation can spread rapidly and become ingrained in the collective consciousness.

One key psychological effect of collective false memories is the reinforcement of these inaccuracies through repeated exposure. When individuals hear the same incorrect information multiple times from different sources, they are more likely to accept it as true. This phenomenon, known as the "illusory truth effect," suggests familiarity can breed acceptance. In the context of the Mandela Effect, hearing a misquoted movie line or seeing a misspelled brand name repeatedly can lead to a widespread belief in its accuracy, even among those who initially had no memory of it. The more people discuss and confirm these false memories, the stronger and more pervasive they become.

Social conformity also plays a significant role in the persistence of collective false memories. Humans have an inherent tendency to conform to the beliefs and behaviors of their social group to maintain harmony and avoid conflict. When individuals realize a majority of people around them share a particular memory, they are likely to adopt that memory themselves, even if it contradicts their own initial recollection. This desire for social cohesion can lead to the suppression of doubts and the acceptance of the collective memory, further entrenching the falsehood. Over time, this conformity can transform an individual's perception, aligning it with the group's shared, albeit incorrect, memory.

The psychological impact of collective false memories is further amplified by the role of authority and credibility in shaping beliefs. When false memories are propagated by authoritative or credible sources, such as trusted media outlets, celebrities, licensed merchandise, or influential community leaders, they gain additional legitimacy. People are more inclined to accept and internalize information from sources they consider reliable. In the case of the Mandela Effect, if a widely respected figure, licensed merchandise or media platform reinforces a false memory, it can significantly bolster the collective belief in that memory, making it even harder to dispel.

Moreover, the emotional resonance of shared memories can make collective false memories particularly compelling. Events or details that evoke strong emotions are more likely to be remembered and discussed. The Mandela Effect often involves emotionally charged subjects, such as childhood memories, beloved cultural icons, or significant historical events. These emotionally laden memories are more susceptible to social influence and more likely to be reinforced within a community. The shared emotional experience can create a powerful bond among individuals, further solidifying the collective false memory.

The psychological effects of collective false memories and social influence may be central to understanding the Mandela Effect. Repeated exposure, social conformity, the influence of authority, and emotional resonance all contribute to the formation and persistence of these shared inaccuracies. These factors highlight the complex interplay between individual memory and social dynamics, demonstrating how our recollections are not only personal but also deeply embedded in our social and cultural contexts. Recognizing the power of social influence on memory can help us better understand the nature of collective false memories and the ways in which they shape our perception of reality.

Confabulation

Confabulation is a psychological phenomenon where individuals create false memories without the intention to deceive. This process involves the brain filling in gaps in memory with fabricated details, making these false memories seem real and coherent. Confabulation often occurs when the brain attempts to make sense of incomplete or fragmented information, leading to the creation of plausible but incorrect narratives. This phenomenon can significantly contribute to the Mandela Effect, where large groups of people collectively misremember specific details or events.

One of the primary reasons confabulation occurs is the brain's need for coherence and continuity in our memories. When there are gaps or inconsistencies in our recollections, the brain automatically fills these voids with information that seems logical and fits the existing narrative. This can result in the creation of entirely new memories that feel as real as genuine ones. In the context of the Mandela Effect, confabulation can lead individuals to recall details, such as the spelling of a brand name or the lines from a movie, in a way that fits their preconceived notions or the general consensus, even if these details are incorrect.

Confabulation could also explain the Mandela Effect surrounding Alexander Hamilton and the misbelief he was a U.S. President. Most people are unable to recall every single President the United States has ever had (gap in memory) and then combine this with the information that is out there about Hamilton they do recall (Founding Father and featured on U.S. currency, which are both traits of other U.S. Presidents), the brain then fills in the gap with putting Hamilton in one of the presidential slots.

The influence of confabulation on collective memory can be exacerbated by social and cultural factors. When people discuss their memories with others, they often unconsciously adopt elements of their peers' recollections, incorporating these details into their own memories. This social sharing of information can reinforce and amplify confabulated memories, leading to a shared false memory among a large group of people. For instance, if a group of friends all misremembers a line from a famous movie, their discussions can solidify this incorrect memory, making it seem more accurate over time. This collective reinforcement is a key element of the Mandela Effect, where widespread confabulation results in a shared, yet incorrect, memory.

Confabulation can also be influenced by external stimuli, such as media representations and cultural references. When people are repeatedly exposed to incorrect information through movies, television, or social media, their brains may incorporate these details into their own memories. For example, a frequently misquoted line from a film can become so ingrained in popular culture that individuals begin to remember it as the actual line, despite it being incorrect. The brain's tendency to confabulate can cause these false memories to feel authentic, leading to a widespread belief in the incorrect information.

Moreover, the emotional and cognitive biases inherent in human memory play a significant role in confabulation. Memories associated with strong emotions or significant life events are more likely to be confabulated, as the brain strives to create a coherent and meaningful narrative. The Mandela Effect often involves memories tied to nostalgic or emotionally charged subjects, making them particularly susceptible to confabulation. People may also be more prone to confabulation when they are confident in their knowledge or when the false memory aligns with their beliefs and expectations.

In conclusion, confabulation is a crucial psychological mechanism that can possibly explain the Mandela Effect by causing individuals to create false memories that feel genuine. The brain's need for coherence, the influence of social and cultural factors, exposure to incorrect information, and emotional biases all play into the formation and reinforcement of these false memories. Understanding the role of confabulation in memory can help explain why so many people share the same incorrect recollections, highlighting the complex and often fallible nature of human memory.

Misleading Post-Event Information

Misleading post-event information may be a significant factor in the formation of false memories, contributing to phenomena like the Mandela Effect. This occurs when people receive incorrect or misleading information after an event, which subsequently alters their memory of the original event. This misinformation can come from various sources, such as conversations, media reports, or even subtle suggestions, leading individuals to incorporate these inaccuracies into their recollections. The altered memories often feel as real and accurate as true memories, highlighting the malleable nature of human memory.[lx]

One of the most well-documented examples of misleading post-event information is the misinformation effect, a phenomenon where people's memories become less accurate due to exposure to incorrect information after the fact. Research by psychologists like Elizabeth Loftus has shown even subtle cues or leading questions can significantly distort a person's recollection of an event. For instance, if someone is asked how fast cars were going when they "smashed" into each other versus when they "hit" each other, they are likely to remember the crash as more severe in the first scenario.[lxi] This demonstrates how powerful and influential misleading information can be on memory.

In the context of the Mandela Effect, misleading post-event information can spread through various channels, leading to collective misremembering. Media plays a crucial role in this process, as incorrect information can be disseminated widely and rapidly. When movies, television shows, or news reports contain inaccuracies, these errors can become embedded in the public's memory. For example, if a popular television show incorrectly references a historical event or a famous quote, viewers might adopt this incorrect version as their own memory. Over time, repeated exposure to the same misinformation solidifies these false memories, making them harder to distinguish from accurate ones.

As with other possible psychological effects, social interactions further amplify the impact of misleading post-event information. When people discuss their memories with others, they are often influenced by the recollections and interpretations of those around them. If a significant number of people share the same incorrect information, it can create a powerful consensus effect, where individuals conform to the group's memory despite their initial doubts. This social validation of false memories can make them feel more credible and accurate, reinforcing the Mandela Effect. In this way, misleading information can propagate through communities, leading to widespread and persistent false memories.

The psychological mechanisms underlying the influence of misleading post-event information also involve cognitive biases. One such bias is the confirmation bias, where people tend to seek out and accept information that confirms their preexisting beliefs while disregarding contradictory evidence. When individuals encounter misleading information that aligns with their expectations or prior knowledge, they are more likely to incorporate it into their memory. This selective acceptance of information can reinforce and perpetuate false memories, making them more resistant to correction.

Misleading post-event information is a key factor in the formation and persistence of false memories, contributing to the Mandela Effect. The misinformation effect, the role of media, social interactions, and cognitive biases all play significant roles in how misleading information can alter our memories. Understanding these mechanisms highlights the vulnerability of human memory to distortion and underscores the importance of critical thinking and verification in preserving accurate recollections. The Mandela Effect serves as a reminder of the intricate interplay between memory, information, and social influence in shaping our perception of reality.

Priming

Priming occurs when exposure to one stimulus influences the response to a subsequent stimulus, often without conscious guidance or intention. This effect can subtly shape perceptions, thoughts, and behaviors, including the formation and recollection of memories. In the context of the Mandela Effect, priming can lead individuals to misremember details or events by predisposing their minds to certain associations or expectations. When people are primed with specific information, they are more likely to recall related but potentially inaccurate details, contributing to the collective false memories characteristic of the Mandela Effect.

One way priming influences memory is through semantic associations. When individuals are repeatedly exposed to a particular piece of information, their brains create connections between that information and related concepts. For example, if someone frequently hears the incorrect version of a famous movie quote, such as "Luke, I am your father" from Star Wars, they may form a strong association between the character and the misquoted line. This repeated priming reinforces the incorrect memory, making it more likely to be recalled in place of the actual line, "No, I am your father." Over time, the primed incorrect information can become so ingrained it feels more accurate than the true memory.

Media and cultural references are powerful sources of priming that can shape collective memories. Advertisements, movies, television shows, memes, and social media can all prime individuals with specific details which may not be accurate. When these primed details are widely disseminated and repeatedly encountered, they can create a shared false memory among large groups of people. For instance, if a popular television show, like *The Big Bang Theory* repeatedly references the Berenstain Bears as *Berenstein Bears*, viewers are primed to remember the incorrect spelling/pronunciation. This widespread priming effect can lead to a collective misremembering, a hallmark of the Mandela Effect.

Social interactions further amplify the impact of priming on memory. When people discuss their memories with others, they can unconsciously prime each other with specific details or interpretations. This social priming can reinforce and spread false memories within a group. For example, if several people in a conversation mention remembering Nelson Mandela's death in the 1980s, this primes others to recall similar false memories, even if they initially had no recollection of such an event. The social validation of these primed memories strengthens the collective belief, contributing to the Mandela Effect.

Cognitive biases also play a role in how priming affects memory. Confirmation bias, the tendency to favor information that confirms preexisting beliefs, can interact with priming to reinforce false memories. When individuals are primed with information that aligns with their expectations or beliefs, they are more likely to accept and remember it, even if it is incorrect. For example, if someone believes a certain historical figure, like Hamilton, was a U.S. President and is repeatedly primed with incorrect references to this effect, they are more likely to recall and believe the false memory. This interaction between priming and cognitive biases underscores the complex mechanisms behind the Mandela Effect.

Through semantic associations, media influence, social interactions, and cognitive biases, priming can predispose individuals to recall inaccurate details or events. Understanding the role of priming in memory distortion highlights the malleability of human recollection and the intricate interplay between external stimuli and internal cognitive processes. Recognizing the impact of priming can help us better appreciate the dynamics of collective false memories and the importance of critical thinking in evaluating our perceptions of reality.

Social Influence and Collective False Memories

Social influence can play a crucial role in the formation and reinforcement of collective false memories. This psychological phenomenon occurs when individuals conform to the memories and beliefs of a group, even when those memories are incorrect. Social influence can manifest in various forms, such as peer pressure, media exposure, and cultural norms. When a significant number of people recall an event or detail inaccurately, their collective agreement can convince others to adopt the same false memory, thereby creating a widely accepted but incorrect version of reality.

The interplay between social influence, priming, and confabulation highlights the complexity of memory formation and distortion. Social influence primes individuals with specific information, which can then lead to confabulated memories that fit the group's narrative. Once these false memories are established, they become resistant to correction due to the reinforcing nature of social interactions. This cycle perpetuates the Mandela Effect, making it challenging for individuals to distinguish between accurate and inaccurate memories. Understanding these psychological mechanisms underscores the importance of critical thinking and skepticism in evaluating collective memories and the information we encounter daily.

Cognitive Dissonance and the Need for Consistency

Cognitive dissonance is a psychological phenomenon where individuals experience discomfort when they hold two or more conflicting beliefs, values, or attitudes. It's that ick feeling these Mandela Effects can give us, when we feel something is off with the world. This discomfort often motivates people to reduce the inconsistency, either by changing their beliefs or by rationalizing the conflicting information. In the context of the Mandela Effect, cognitive dissonance can also have the potential to cause us to misremember something. It can lead us to misremember details to align with the consensus of a group. When confronted with evidence that contradicts their memories, individuals may unconsciously alter their recollections to reduce the dissonance and maintain internal consistency, even if the evidence was incorrect.

The need for consistency is a powerful driver of human behavior and cognition. People have a natural inclination to create a coherent and stable understanding of the world around them. When they encounter discrepancies between their memories and external information, the desire for consistency can lead to the alteration of memories. For example, if someone strongly believes they have always known a particular brand name to be spelled a certain way, encountering the correct spelling can create cognitive dissonance. To resolve this, they might misremember the original spelling to match their belief, thus experiencing a Mandela Effect.

This need for consistency is also influenced by social factors. When individuals are part of a group that collectively remembers an event or detail incorrectly, the pressure to conform can intensify cognitive dissonance. Aligning one's memories with the group's consensus reduces social tension and reinforces a sense of belonging. As a result, individuals may unconsciously adopt the group's false memories to maintain social harmony and personal consistency. This process explains how collective false memories, or Mandela Effects, can become widespread and deeply ingrained in a community.

Cognitive dissonance also interacts with other psychological mechanisms, such as confabulation and priming. When people are primed with incorrect information, their need for consistency can lead them to confabulate details that support the primed memory. For instance, if a person is repeatedly exposed to a misquoted line from a famous movie, their cognitive dissonance may drive them to remember the incorrect version to align with the repeated exposure. This interplay between cognitive dissonance, priming, and confabulation highlights the complexity of memory distortion and the formation of Mandela Effects.

Understanding cognitive dissonance and the need for consistency provides insight into why people might misremember details and events. The discomfort of holding conflicting beliefs and the motivation to maintain a stable worldview drive individuals to unconsciously alter their memories. When combined with social influence and other cognitive biases, these processes could be part of the reason for the widespread phenomenon of the Mandela Effect, where large groups of people share the same incorrect memories.

Memory Reconstruction and Schemas

Memory reconstruction and schemas play a crucial role in how people remember and misremember events, which could contribute to phenomena like the Mandela Effect. Memory reconstruction refers to the process by which memories are pieced together from various bits of information rather than being recalled as exact replicas of past events. This process is inherently fallible, as it relies heavily on the individual's cognitive frameworks, or schemas. Schemas are mental structures that help organize and interpret information based on past experiences and knowledge.

Schemas can lead to misremembering because they fill in gaps in our memories with what we expect to have happened, rather than what actually did. For instance, if a person has a schema a famous quote is worded in a particular way, their mind might reconstruct the memory to fit this schema, even if it's incorrect. This is why so many people remember the line "Luke, I am your father" from Star Wars, despite the actual line being, "No, I am your father." The schema for how famous movie quotes are typically remembered can override the accurate details.

Furthermore, when people share their reconstructed memories with others, these memories can influence the schemas of those around them, leading to a collective misremembering. This social reinforcement solidifies the false memories, making them more resistant to correction. The more frequently a reconstructed memory is recalled and shared, the more entrenched it becomes, leading to widespread acceptance of inaccuracies, as seen in various Mandela Effects.

This process is exacerbated by the fact each time a memory is recalled, it is susceptible to being altered by current perceptions, emotions, and additional information. The malleability of memory means what started as a small error in recall can snowball into a widely accepted false memory. Understanding memory reconstruction and schemas sheds light on why people might misremember events or details, illustrating the powerful influence of cognitive frameworks and social dynamics in shaping our recollections.

CHAPTER 6:
Psychological Explanations for the Mandela Effect

Conspiracy theory-type reasons for the Mandela Effect suggest more extraordinary explanations, often invoking ideas beyond conventional understanding. Some believe the Mandela Effect is evidence of parallel universes or alternate realities intersecting with our own, causing discrepancies in collective memories. Others propose these effects result from "glitches in the matrix," implying our reality is a simulated environment subject to occasional errors. Additionally, there are theories about intentional manipulation of information by powerful entities to test or control public perception and memory. These speculative ideas offer intriguing, albeit unproven, alternatives to psychological explanations, so let's take a look at them.

Parallel Universes

Parallel Universes Overview

Parallel universes, also known as alternate universes or multiverses, are theoretical realms theorized to exist alongside our own universe but operate independently with their own distinct laws of physics and realities. The concept suggests there are an infinite number of universes, each differing from the other in various ways, from minor variations to entirely different sets of physical laws. This idea stems from various fields, including cosmology, quantum mechanics, and string theory, and has been popularized by science fiction.

One interpretation of alternate universes arises from the Many-Worlds Interpretation of quantum mechanics. Proposed by physicist Hugh Everett in 1957, this theory suggests every quantum event branches off into a separate universe, creating a vast, possibly infinite, number of parallel realities. Each universe represents a different outcome of these quantum events, leading to a multiverse where every possible history and future exists simultaneously[lxii].

Another perspective comes from cosmology and string theory, where the idea of a multiverse emerges from the concept of an inflationary universe. According to this theory, the rapid expansion of the universe after the Big Bang could have led to the formation of numerous isolated regions, or "bubble universes," each with its own distinct properties. These bubble universes are thought to be part of a larger multiverse, existing beyond the observable universe[lxiii].

The concept of alternate universes also raises fascinating philosophical and existential questions. It challenges our understanding of reality, suggesting our universe might be just one of countless others. This idea has captivated the imagination of both scientists and the public, providing a rich ground for speculation, research, and storytelling in science fiction. While the existence of alternate universes remains theoretical and unproven, it continues to be a topic of intense debate and exploration in modern physics and cosmology.

Parallel Universes and the Mandela Effect

Parallel universes may offer a fascinating explanation for the Mandela Effect, suggesting our reality is just one of many. According to the Many-Worlds Interpretation of quantum mechanics, every possible outcome of a quantum event results in the creation of a new, parallel universe[lxiv]. This means countless alternate versions of reality could exist simultaneously, each with its own unique history and events. When people experience the Mandela Effect, they might be tapping into memories from these parallel universes, where events happened differently than in our own.

The idea of parallel universes aligns with the concept of memory reconstruction, where our brains fill in gaps in our recollections based on existing schemas and new information. If alternate realities do exist, individuals might occasionally access these different versions of events, leading to widespread but incorrect memories. For instance, the widespread belief Nelson Mandela died in prison in the 1980s could be a memory from a parallel universe where this event actually occurred, causing confusion and the Mandela Effect when individuals share these conflicting recollections.

Moreover, the theory of parallel universes challenges our understanding of reality and consciousness. It raises the possibility our perceptions and memories are not confined to a single, linear timeline but are instead influenced by multiple, coexisting realities. This concept provides a compelling explanation for why large groups of people might share the same false memories, as it suggests we might be interconnected with other versions of ourselves in alternate universes. While the theory remains speculative and unproven, it offers a rich area for exploration in both science and philosophy, pushing the boundaries of our understanding of memory and existence.

Overall, the idea of parallel universes as an explanation for the Mandela Effect underscores the complexity of human cognition and the potential for reality to be far more intricate than we currently comprehend. It invites us to consider the limits of our knowledge and the possibility our memories might be influenced by experiences beyond our immediate reality. While it remains a topic of debate and exploration, the multiverse theory continues to captivate the imagination and offers a profound perspective on the nature of memory and the Mandela Effect.

Alternate Realities

Alternate Realities Overview

Although some may use the terms parallel universes and alternate realities interchangeably, but there are nuanced differences between the two. Alternate realities refer to worlds that exist due to divergence from our own. Think of them as forks in the road of a single universe. Each decision (such as what if I didn't include this explanation of the two terms in this book) forking off the reality timeline we currently are in (where I do indeed include this section). Everything prior to that fork in the road between the main reality and the alternate reality are exactly the same. In contrast, parallel universes exist independently. Each parallel universe has its own unique laws of physics and unique history. These laws and history may be similar in many ways to our own, but (normally) nothing we do in this universe affects the parallel universe. Going back to the road analogy, think of it as we're on a highway, and the parallel universe is a frontage road running alongside of us. The frontage road has a different speed limit, likely a different number of lanes of traffic. They may have stop signs and may be able to directly access businesses we can't even get to from our highway.

Alternate Realities and the Mandela Effect

One captivating explanation for the Mandela Effect is the theory of alternate realities, where different versions of events unfold as our reality forks away from others. Perhaps it was an alternate reality where the

Bearenstains came to America and their name was misspelled Bearenstein when they arrived, and it stuck. According to this theory, individuals who experience the Mandela Effect may be accessing memories from alternate realities where these events happened differently. In some theories, it is theorized those who are experiencing the Mandela Effect are actually from an alternate reality and have somehow been transported into this reality, without their knowing. This hypothesis opens a fascinating avenue for understanding the complexities of memory and perception.

But wouldn't we know if we were from an alternate reality?

Not necessarily.

There's a brilliant blogger, Reece, who lightly explains the mathematics behind the physics of alternate realities. Using imaginary numbers, the theories of relativity by Einstein and Minkowski, and quantum field theory, he theorizes alternate realities could just be different hexadectants of the same universe. Using the BearenstEin and BearenstAin aspects of one of the most popularly known Mandela Effects to signify two different universes, he explains:

> I further propose that the stEin and the stAin universes are actually just different hexadectants of the same universe: in the stEin universe, all three spatial dimensions are real and time is imaginary; in the stAin universe, all three spatial dimensions are imaginary and time is real. Of course, from the standpoint of stEin/stAin this won't produce any mathematically significant difference; it's the same as choosing (+++-) or (---+) convention for Minkowski space, which at the end doesn't alter predictions or measurements. We'd never know if we did swap[lxv].

The idea of alternate realities challenges our traditional understanding of time and space. It suggests our linear perception of history may be just one of many possible paths and other versions of events could coexist alongside our own. This perspective not only provides a potential explanation for the Mandela Effect but also invites us to reconsider the nature of reality itself. The existence of alternate realities would mean the universe is far more complex and interconnected than we currently understand, with profound implications for science and philosophy.

Reality is a Simulation

Reality is a Simulation Overview

The conspiracy theory that surmises our reality is a computer simulation, similar to the movie "The Matrix," suggests our perceived world is actually a sophisticated artificial construct. This theory gained traction from the philosophical proposition by Nick Bostrom, who argued future civilizations could develop technology capable of creating such simulations[lxvi]. According to this idea, if it were possible to simulate entire realities, it is statistically likely our reality could be one of these simulations.

In the popular 1999 movie *The Matrix*, characters live in a simulated world designed to keep them unaware of their true existence, controlled by external forces. Similarly, proponents of the simulation hypothesis believe our everyday experiences, memories, and perceptions might be artificially generated. This concept challenges fundamental notions of existence and reality, suggesting everything we know might be part of a complex computational program.

Supporters of this theory point to various scientific and philosophical arguments. For instance, the rapid advancement of virtual reality and artificial intelligence demonstrates the potential for creating immersive simulated environments. While highly speculative and lacking empirical evidence, the idea of living in a computer simulation continues to fascinate and provoke debate among scientists, philosophers, and the public.

Glitches in the Simulation as an Explanation for Mandela Effects

One of the more intriguing theories regarding the Mandela Effect centers on our reality actually being some sort of simulation. The Mandela Effects we experience, under this hypothesis, can be explained as glitches or anomalies in the simulation. These glitches result in inconsistencies or changes in our collective memories, leading to widespread false recollections of events, names, or details.

In this context, the Mandela Effect represents moments when the underlying code of the simulation is altered or experiences errors. These alterations could be minor, like the spelling of a word, or more significant, like the memory of a historical event. Such glitches could explain why large groups of people remember events differently from documented history. If we are living in such a simulation, the Mandela Effect could be evidence of the limitations or imperfections of this artificial construct. It challenges our understanding of reality, suggesting our perceptions and memories might be influenced by factors beyond our awareness.

While the simulation theory remains speculative and controversial, it offers a thought-provoking explanation for the Mandela Effect. It integrates elements of science fiction, philosophy, and advanced technology to propose a radical rethinking of reality. Although there is no empirical evidence to support the idea we live in a simulation, it continues to capture the imagination and spark debate, highlighting the profound mysteries of human memory and the nature of existence.

Powerful Entities Controlling Our Perception

Powerful Entities Controlling Our Perception Overview

The conspiracy theory that powerful entities, such as governments, big businesses, and shadow organizations, manipulate society is rooted in the belief these groups control significant aspects of global affairs. Proponents of this theory argue these entities wield immense power behind the scenes, influencing political decisions, economic policies, and social movements to serve their interests rather than the public good. This notion is often fueled by a lack of transparency in governmental and corporate actions, leading people to suspect there is more happening behind closed doors than is publicly acknowledged[lxvii].

Governments

One of the central arguments of this theory is governments engage in covert operations and propaganda to maintain control over the populace. This includes allegations of surveillance programs, psychological operations, and disinformation campaigns designed to keep the public unaware of the true extent of their power. Historical events such as the Watergate scandal and the revelations about the NSA's surveillance practices lend some credibility to these concerns, as they demonstrate governments can and do engage in secretive and manipulative behaviors. Projects like the CIA's MKUltra project which involved the use of hallucinogens as not only a means of interrogation, but in an effort to produce "Manchurian candidates" are proof positive that governments have (and likely still are) manipulated their citizens' minds.

Big Business

Big businesses, particularly multinational corporations, are also accused of manipulating society for profit. Critics argue these corporations use their vast resources to influence political leaders through lobbying and campaign contributions, ensuring laws and regulations favor their interests. This can lead to policies that prioritize corporate profits over public welfare, such as deregulation of industries, tax breaks for the wealthy, and suppression of labor rights. The 2008 financial crisis, where major financial institutions were seen as playing a key role in economic collapse while facing minimal repercussions, is often cited as an example of corporate manipulation.

Shadow Organizations

Shadow organizations, often referred to as "deep state" or "secret societies," are believed to be the hidden hands guiding global events. These groups are thought to operate beyond the reach of democratic institutions, orchestrating major political and social changes. Theories about groups like the Illuminati, Freemasons, and Bilderberg Group are prevalent in this context, suggesting a small, elite group of individuals holds disproportionate influence over world affairs[lxviii]. While evidence for the existence and influence of such groups is largely anecdotal and speculative, the idea persists in popular culture and conspiracy lore.

These conspiracy theories thrive in part due to the complex and often opaque nature of modern global systems, where the lines between public and private interests can be blurred. They also reflect a broader skepticism and mistrust of authority figures and institutions, which is exacerbated by instances of genuine corruption and abuse of power[lxix]. While it is important to question and hold powerful entities accountable, the proliferation of conspiracy theories can also lead to misinformation and a distorted understanding of how power operates in society.

Powerful Entities and Mandela Effects

Another theoretical explanation for Mandela Effects involves powerful entities and their desire to manipulate society through the creation of these Mandela Effects. It suggests governments, corporations, and shadow organizations intentionally alter reality to control public perception. According to this view, these entities possess advanced technology or knowledge that allows them to subtly change details in our shared reality, creating false memories among large groups of people. By doing so, they can confuse, distract, or mislead the public, preventing cohesive resistance and maintaining their dominance over societal structures.

One argument supporting this theory is the potential use of psychological operations and propaganda to influence collective memory. Governments have historically engaged in disinformation campaigns to control public opinion, and the concept of the Mandela Effect could be an extension of these tactics. By subtly altering facts and creating confusion, these powerful entities can undermine the public's trust in their own memories and in reliable sources of information, making it easier to push specific agendas[lxx].

Corporations, particularly those in the tech industry, are also implicated in this conspiracy theory. With their vast resources and control over digital media, these companies could theoretically alter content across the internet, contributing to the Mandela Effect. For instance, they could change search engine results, manipulate social media algorithms, or even edit digital archives to create false memories. This manipulation would serve their interests by shaping consumer behavior and perceptions of reality in ways that benefit their business goals.

Shadow organizations too are suspected of orchestrating Mandela Effects. These entities are believed to operate behind the scenes, directing major political and social changes without public knowledge. Theories about these groups suggest they alter reality, resulting in Mandela Effects, which could be one of their tools to maintain control and further their hidden agendas[lxxi].

While there is no concrete evidence to support these claims, the idea powerful entities could be causing Mandela Effects reflects broader societal concerns about the transparency and accountability of those in power. It underscores a deep-seated mistrust of authority and the fear our perceptions and memories can be manipulated by forces beyond our control. This theory, while speculative, highlights the importance of critically examining the sources and influences that shape our understanding of reality.

CHAPTER 7:
The Internet's Impact on Mandela Effects

The internet has fundamentally transformed how information is shared and consumed, impacting every facet of modern life, including our collective memory. As a powerful tool for communication, it has the capability to rapidly disseminate information across the globe, reaching millions of people within seconds. This unprecedented connectivity has led to the formation of global communities and the rapid spread of ideas, including those related to the Mandela Effect.

The internet's role in shaping and reinforcing the Mandela Effect cannot be overstated. Social media platforms, online forums, and digital archives serve as both repositories and amplifiers of these collective false memories. When a particular memory discrepancy is shared online, it often gains traction as more people contribute their own similar experiences, leading to widespread belief in the altered memory. This process is further compounded by the algorithms that govern social media and search engines, which prioritize content based on popularity and engagement rather than accuracy.

Moreover, the internet provides a fertile ground for the proliferation of misinformation and fake news, which can distort public perception and contribute to the creation of false memories. In the context of the Mandela Effect, misinformation can spread rapidly through viral content, memes, and influential online personalities, further embedding these erroneous recollections in the collective consciousness. As users engage with this content, they become part of a feedback loop that reinforces the incorrect memories, making them more resistant to correction.

Let's explore the various ways the internet influences the Mandela Effect, examining the roles of social media, digital echo chambers, memes, misinformation, influencers, and online discussions. By understanding these dynamics, we can gain insight into how collective memories are shaped in the digital age and the implications for our perception of reality.

Social Media Amplification

Social media platforms like Facebook, Twitter, and Reddit play a significant role in amplifying Mandela Effects. When a user posts about a memory discrepancy, such as a misremembered detail from a movie or book, it can quickly go viral if others resonate with the experience. The interactive nature of social media encourages users to share their own similar experiences, creating a snowball effect that spreads the false memory to a larger audience.

Algorithms on these platforms further enhance the spread of Mandela Effects by prioritizing engaging content. Posts that generate a lot of comments, likes, and shares are more likely to appear in users' feeds, regardless of their accuracy. This engagement-based promotion can lead to widespread belief in incorrect memories as more people see and interact with the content, reinforcing the false memory through repeated exposure.

Digital Echo Chambers

Digital echo chambers are environments, particularly online, where individuals are predominantly exposed to information, ideas, or opinions that align with their existing beliefs. These echo chambers occur on social media platforms, forums, and other online communities where algorithms and user behavior create a feedback loop, reinforcing users' preconceptions and filtering out contradictory viewpoints. This selective exposure can lead to the strengthening of personal biases and the spread of misinformation, as users primarily interact with content and people who validate their own perspectives, often resulting in a skewed understanding of reality.

These can play a crucial role in reinforcing and spreading Mandela Effects by creating environments where users are primarily exposed to information that confirms their existing beliefs and memories. Within these echo chambers, people with similar experiences and perspectives share and validate each other's memories, including those that are false.

This mutual reinforcement can make the false memories more robust and less susceptible to correction, as individuals become more confident in the accuracy of their recollections due to the collective validation they receive.

The algorithms that drive content on social media platforms and forums often prioritize posts that match a user's interests and past interactions, further intensifying the echo chamber effect. When users engage with content related to a specific Mandela Effect, the platform's algorithms are likely to show them more similar content, creating a feedback loop. This selective exposure means users are repeatedly confronted with information that aligns with their false memories, making it harder for contradictory evidence to penetrate their belief system.

Online communities and forums dedicated to Mandela Effects serve as prime examples of digital echo chambers. These spaces allow individuals to share their experiences and discuss various instances of the Mandela Effect, often reinforcing each other's false memories. As these communities grow, the collective belief in the accuracy of these altered memories strengthens, with members frequently dismissing contradictory evidence as inaccurate or irrelevant. This communal validation can solidify false memories, making them a significant part of the collective consciousness.

The nature of echo chambers also means misinformation can spread rapidly and widely within these groups. Once a false memory gains traction, it is often repeated and accepted without critical examination, as members of the echo chamber trust the information shared by their peers. This can lead to widespread acceptance of incorrect details and further propagation of Mandela Effects. In this way, digital echo chambers not only reinforce false memories but also contribute to their initial formation and dissemination.

Memes and Viral Content

Memes and viral content may also play a significant role in the propagation of Mandela Effects. Memes, by nature, are designed to be easily shareable and relatable, often distilling complex ideas into simple, humorous images or phrases. When a meme encapsulates a misremembered fact or false memory, it can rapidly spread across the internet, reaching vast audiences who may adopt the false memory as truth. This process is exacerbated by the tendency of memes to be replicated and modified, further embedding the incorrect information into the collective consciousness.

The viral nature of internet content ensures once a Mandela Effect meme gains traction, it can achieve widespread visibility in a very short period. Social media platforms like Facebook, Instagram, and X facilitate this rapid dissemination through likes and shares. As more people interact with and share the meme, the false memory becomes more ingrained and accepted as factual by the public. This repetition and reinforcement can make it challenging for individuals to distinguish between the original memory and the altered one presented by the meme.

Moreover, the humor and relatability of memes contribute to their effectiveness in spreading false memories. When people find a meme funny or relatable, they are more likely to remember and share it. This emotional engagement can make the false memory more memorable than the correct information, especially if the meme aligns with pre-existing beliefs or schemas. Consequently, the humor and brevity of memes can significantly influence how people recall and interpret past events, contributing to the persistence and spread of Mandela Effects.

Misinformation and Fake News

Misinformation and fake news are critical factors in the spread of Mandela Effects. Fake news refers to false or misleading information presented as legitimate news, often to deceive or manipulate public perception. In the context of Mandela Effects, fake news can plant seeds of false memories, which can then spread through social media and other online platforms. When people encounter these fabricated stories, they may internalize the incorrect details, contributing to the formation and reinforcement of Mandela Effects.

Clickbait headlines and sensational stories often garner more attention and shares than fact-checked news, leading to widespread dissemination of false information. When individuals encounter these misleading or incorrect reports, they may incorporate the false details into their memories, thus reinforcing the Mandela Effect phenomenon. This dynamic is particularly powerful when the misinformation aligns with pre-existing beliefs or biases, making the false memories more persistent.

The phenomenon of fake news can be exacerbated by confirmation bias, where individuals favor information that confirms their existing beliefs and ignore contradictory evidence. This bias can lead people to accept and spread fake news that supports their misconceptions, further entrenching the Mandela Effects in collective memory[lxxii]. As people encounter and share this misinformation within their social networks, the false memories become more deeply embedded and harder to dislodge, creating a cycle of reinforcement that perpetuates the Mandela Effect.

The Role of Influencers and Celebrities

Influencers and celebrities significantly impact the spread of Mandela Effects due to their extensive reach and the trust their followers place in them. Internet personalities, YouTubers, and social media influencers often discuss and share their own experiences with Mandela Effects, further popularizing these phenomena. When a well-known figure endorses a false memory, it can quickly gain credibility and spread to a larger audience, making the Mandela Effect more pervasive.

Case studies highlight instances where influencers have perpetuated false information and how it becomes widely accepted as truth[lxxiii]. This can translate to false memories as well. For example, popular YouTubers might create videos discussing their confusion about a specific Mandela Effect, such as the Berenstain Bears spelling or the color of C-3PO's leg in Star Wars. These videos often receive millions of views, and the comments section typically fills with viewers expressing their agreement and sharing their similar experiences. This collective reinforcement solidifies the false memories, making them more resistant to correction.

The reach and impact of influencers and celebrities on spreading Mandela Effects cannot be underestimated. Their large followings and frequent interactions with fans ensures any content they produce has the potential to go viral[lxxiv]. The combination of their authoritative voice and the platform's algorithm, which promotes engaging content, amplifies their message. This dynamic creates a powerful feedback loop where false memories are continuously shared, discussed, and reinforced across various social media channels.

Online Debates and Discussions

Online debates and discussions play a significant role in the spread and reinforcement of Mandela Effects, beyond just what information is presented to us. Unlike communication methods of just a generation ago, where the information only flowed in one direction (such as television newscasts, radio, and newspapers), today we have forums, comment sections, and social media platforms where we can share our experiences and perceptions of these phenomena, making these communications interactive discussions, not only with the presenter of the information, but also others from around the world. These interactions often lead to memory distortion, as repeated exposure to others' false memories can influence and alter an individual's recollection of events. The collaborative nature of these discussions creates a feedback loop, where shared inaccuracies are validated and reinforced within the group.

The psychological effects of group discussions on individual memories are profound. Groupthink, a psychological phenomenon where the desire for harmony and conformity within a group leads to irrational decision-making, can play a significant role in memory distortion[lxxv]. When individuals participate in discussions about Mandela Effects, the consensus within the group can pressure members to conform to the shared false memories. This social pressure can cause individuals to doubt their original memories and adopt the group's incorrect version, further perpetuating the Mandela Effect.

Internet Archives and Search Engines

Internet archives, such as the Wayback Machine, play a critical role in verifying or refuting Mandela Effects. These archives preserve snapshots of web pages over time, allowing users to track changes and confirm historical data. For instance, users can investigate the spelling of a name by accessing archived versions of the official website from different years. This capability helps clarify the accuracy of collective memories by providing concrete evidence (Pennycook & Rand, 2019).

However, it's important to remember the internet only became publicly available in the early 1990s. Many Mandela Effects involve things that are misremembered from much earlier. In an instance like the Bearenstain vs Bearenstein Bears, although the Wayback Machine can look at websites from years past, the first recorded official website for the Bearenstain Bears was on December 8th, 1998, yet the first book, *The Big Honey Hunt,* was published in 1962. Obviously, internet archives wouldn't prove the Ein version of the name never existed. It only shows the website was spelled Ain. However, if you subscribe to one of the conspiracy theory type explanations, you probably won't trust a website to accurately report historical data. Plus, even if we assume a site like Wayback Machine is 100% accurate in what it presents for website data, it still doesn't give any evidence, either way, for books or even the TV series. This dual nature of digital archives underscores the importance of critical evaluation when using them to investigate Mandela Effects[lxxvi].

Search engines can significantly influence the dissemination of information and its accuracy. They often prioritize popular or highly engaging content, which can include misinformation or perpetuated false memories. When individuals search for information about Mandela Effects, the results they encounter can shape their beliefs and reinforce existing misconceptions. The algorithmic nature of search engines can sometimes lead to an echo chamber effect, where users are repeatedly exposed to similar information, further entrenching false memories (Allcott & Gentzkow, 2017).

Conclusion

As we conclude this exploration of the Mandela Effect, it's evident this fascinating phenomenon touches on various aspects of human cognition, perception, and societal influence. From the initial overview and definition of the Mandela Effect, we delved into how cognitive biases shape our memories, often leading us to remember events and details inaccurately. Understanding memory formation and retrieval, along with factors influencing memory accuracy, highlighted the complexities of human memory and the role of perception in shaping our reality.

We examined some of the most famous Mandela Effects, showcasing how widespread and diverse these memory anomalies can be. Whether it's the perplexing case of the Berenstain Bears' spelling, the misremembered death of Nelson Mandela, C-3PO's silver leg, or the mistaken recollections of movie quotes and more, these examples illustrate how our collective memories can diverge from reality in intriguing ways.

Psychological explanations offer a deeper insight into the mechanisms behind the Mandela Effect. Concepts such as collective false memories, confabulation, and priming demonstrate how our minds can be influenced by social interactions and post-event information. The need for cognitive consistency and the reconstruction of memories based on schemas further explained why these memory distortions occur.

Exploring alternate explanations, such as parallel universes, alternate realities, and the simulation hypothesis, provided a thought-provoking perspective on the Mandela Effect. These theories, though more speculative, offer intriguing possibilities for why our memories might not always align with reality. The idea powerful entities could manipulate our perception added another layer of complexity to our understanding of this phenomenon.

Finally, the impact of the internet on the Mandela Effect cannot be overstated. Social media amplification, digital echo chambers, and the rapid spread of memes and viral content all contribute to the persistence and reinforcement of false memories. Influencers and celebrities play a significant role in perpetuating these effects, while online debates and discussions further embed these inaccuracies in our collective consciousness. The availability of internet archives and the role of search engines in disseminating information highlight both the potential for clarifying and complicating our understanding of Mandela Effects.

Yet, despite these logical explanations, there is still that *ick* feeling I personally experience when I think about some of the Mandela Effects I've experienced in my life. I like to think I haven't been so influenced by misinformation or others' opinions. I like to think my memories are my own and correct. To think my memory has been so distorted is disturbing. However, when I consider the more speculative possible reasons why my memory is different from recorded history, well, this is even more disturbing.

> This experience truly does feel like if I just pull on the right thread, I will indeed unravel reality.

There are benefits to exploring Mandela Effects, even if there are no clear-cut answers as to why they happen. Mandela Effects remind us how we should all approach our recollections with a critical and open mind. Our memories are not infallible, and they definitely can be influenced by outside factors.

This exploration also highlights the importance of taking in information with a grain of salt, especially when it comes from the internet, but even when it comes from what should be trusted sources. Remember, sometimes even trusted sources get information wrong or are even tricked into disseminating wrong information.

Consider the "chocolate diet" hoax that occurred in early 2015. Researchers looking to expose pseudoscience published a heavily-biased study saying eating dark chocolate everyday could help people lose weight. They even created a fictitious health organization, the Institute of Diet and Health, to help make the study look more legitimate.

The result?

These fake "results" were published widely across the internet. Even respected periodicals like *Prevention* and *Shape* talked about these exciting (but completely fake) findings[lxxvii]. It served to dramatically point out how quickly and thoroughly misinformation can be spread nowadays. And when this information is incorrect about the past, it could possibly then impact our recall of events, potentially resulting in Mandela Effects.

An honest mistake, meant to be a joke, or for nefarious purposes, the reality is there is misinformation being shared at an exponentially faster rate than ever before in history. For this reason, it's important to remember to do your due diligence when taking in information. Even respected information sources, like the periodicals that picked up the "chocolate diet" news can mistakenly spread false information. Is this misinformation where Mandela effects stem from…

… Or is this misinformation being used
to cover the fact these Mandela effect memories
are actually real?

Endnotes

[i] Battersby, J. D. (1988, August 17). Mandela hospitalized with tuberculosis, lawyer says. *The New York Time*, p. 3.

[ii] Reece. (2014, June 23). *On the Berenstein Bears switcheroo*. On the Berenstein Bears Switcheroo. https://www.woodbetween.world/2014/06/commenting-on-berenstein-bears.html

[iii] Jake Edmiston. (2015, August 12). *Don't be ridiculous, it was always spelled Berenstain Bears, says son of series creators | National Post*. National Post. https://nationalpost.com/entertainment/its-berenstain-like-coffee-stain-or-jello-stain-one-berenstain-bears-author-rejects-parallel-universe-theory

[iv] YouTube. (2010, August 13). *The Berenstain Bear - old theme song*. YouTube. https://www.youtube.com/watch?v=sjXiIZYsGJY&t=23s

[v] Takei, G. (2015, August 5). *George Takei*. Facebook. https://www.facebook.com/georgehtakei/posts/1327077263988390

[vi] Berenstain, S., & Berenstain, J. (1962). *The big honey hunt*. Beginner Books.

[vii] Cooke, H. (2016, October 12). *NZ and the "Mandela Effect": Meet the folks who remember New Zealand being in a different place*. Stuff. https://www.stuff.co.nz/oddstuff/85258775/nz-and-the-mandela-effect-meet-the-folks-who-remember-new-zealand-being-in-a-different-place

[viii] Onlmaps. (2022, September 22). *Maps on the web*. Tumblr. https://mapsontheweb.zoom-

maps.com/post/696086295235198976/percent-of-people-that-trust-climate-scientists

[ix] Murphy, N. (2019, February 8). Ikea mocked for very obvious mistake on £30 world map - can you notice error? *Mirror*. Retrieved August 1, 2024, from https://www.mirror.co.uk/news/uk-news/ikea-being-mocked-very-obvious-13969490.

[x] *Council on Geostrategy on X: In Another Big Win for the UK...* X (formerly Twitter). (2023, March 13). https://x.com/ConGeostrategy/status/1635382259384737808

[xi] May, N. H., & May, 6. (2021, May 6). *Where's New Zealand? world map places country in confusing new location.* NZ Herald. https://www.nzherald.co.nz/lifestyle/wheres-new-zealand-world-map-places-country-in-confusing-new-location/JVEUGF7CSLUUIVCKUOEDHY35JI/

[xii] Worldbreakers. (2018, April 14). *R/retconned - residue monocle 1994 junior monopoly game.* Reddit. https://www.reddit.com/r/Retconned/comments/8cbb7d/residue_monocle_1994_junior_monopoly_game/

[xiii] *955-D9000-19-D9-4760-AA9-E-F34-D349-E2717 hosted at imgbb*. ImgBB. (n.d.). https://ibb.co/pLRTLQM

[xiv] *PBS.twimg*. PBS.twimg.com. (n.d.). https://pbs.twimg.com/media/Cu5MDLXWgAAVWB3.jpg

[xv] Imgur. (n.d.). *Imgur*. https://i.imgur.com/vrLwWm0.jpeg

[xvi] Hallerman, T. (2017, October 6). Lessons learned after equifax hearings. *Atlanta Constitution*, p. 6.

[xvii] Wikimedia Foundation. (2024, April 19). *Mr. Monopoly*. Wikipedia. https://en.wikipedia.org/wiki/Mr._Monopoly

xviii Monopoly. (2016, May 18). *Facebook - Monopoly*. Facebook. https://www.facebook.com/photo/?fbid=10153545232881517&set=a.63345516516

xix Sports Illustrated. (n.d.). https://www.si.com/.image/ar_1:1%2Cc_fill%2Ccs_srgb%2Cfl_progressive%2Cq_auto:good%2Cw_1200/MTkxMzA5MjklNjI1NzA5MTU0/1241733806.jpg

xx aispacioli. (2021). Few knew how to '90s like Sinbad. : Nostalgia - reddit. https://www.reddit.com/r/nostalgia/comments/kghzsa/few_knew_how_to_90s_like_sinbad

xxi Das, A. (2023, July 7). *Fact check: Did fruit of the loom ever have a cornucopia in their logo? viral claim debunked*. Sportskeeda. https://www.sportskeeda.com/pop-culture/fact-check-did-fruit-loom-ever-cornucopia-logo-viral-claim-debunked

xxii *The fruit story*. Fruit of the Loom. (n.d.). https://www.fruit.com/fruit-story

xxiii *Status search SN 73006089*. US Patent and Trademark Office. (n.d.). https://tsdr.uspto.gov/#caseNumber=73006089&caseSearchType=US_APPLICATION&caseType=DEFAULT&searchType=statusSearch

xxiv kurz_prime. (2024, March 10). *Sure*. X. https://twitter.com/Kurz_Prime/status/1766965377408319598/photo/1

xxv Kasprak, A. (2024, March 21). *The Fruit of the Loom logo has never contained a cornucopia, honestly*. Snopes. https://www.snopes.com/fact-check/cornucopia-fruit-of-the-loom/

xxvi Rothstein, A. (n.d.). *And a little child: 1939.* Shorpy.com. https://www.shorpy.com/node/23002

xxvii ExploreRealityAlso. (2022). R/retconned on reddit: Possible residue for cornucopia from 1900s fruit of the loom advert. https://www.reddit.com/r/Retconned/comments/uio0ws/possible_residue_for_cornucopia_from_1900s_fruit/

xxviii Disney, W., et al. (1937). *Snow White and the Seven dwarfs.* United States; Distributed by Buena Vista Film Distribution Co.

xxix IMDb.com. (n.d.). *Star wars: Episode IV - A new hope.* IMDb. https://www.imdb.com/title/tt0076759/mediaviewer/rm1579585280/

xxx IMDb.com. (n.d.-a). *Star wars: Episode IV - A new hope.* IMDb. https://www.imdb.com/title/tt0076759/mediaviewer/rm2482838785/

xxxi IMDb.com. (n.d.-a). *Star wars: Episode IV - A new hope.* IMDb. https://www.imdb.com/title/tt0076759/mediaviewer/rm663501568/

xxxii Galaxy Publications, Paradise Press, & Bunch, H. (1977). *Star Wars Official Poster Monthly.*

xxxiii Guenette, R. (1977, September 16). The Making of Star Wars. broadcast, ABC.

xxxiv Vintagetoyarchive. (2013, January 27). *Vintage toy archive.* VINTAGE TOY ARCHIVE. https://vintagetoyarchive.tumblr.com/post/41563106513/kenner-1977-c-3po-large-size-action-figure

xxxv *Star wars greatest hits c3po action figure.* Amazon.co.uk: Toys & Games. (n.d.). https://www.amazon.co.uk/Star-Wars-Greatest-Action-Figure/dp/B000GKE01O

[xxxvi] Lucasfilm LTD. (2004). *Star wars: The empire strikes back*. Beverly Hills, CA.

[xxxvii] London Films. (1933). *The Private Life of Henry VIII* [Film].

[xxxviii] Deary, T., Tonge, N., & Brown, M. (2007). *Terrible tudors: Horrible histories*. Scholastic.

[xxxix] Holbein the Younger, H. (1537). *Portrait henry viii* Walker Art Gallery, Liverpool, England.

[xl] Warner, J. L., et al. (1942). *Casablanca*. United States; Warner Bros. Pictures, Inc.

[xli] Warner, J. L., et al. (1942). *Casablanca*. United States; Warner Bros. Pictures, Inc.

[xlii] *The first program*. Mister Rogers' Neighborhood. (2018, September 7). https://misterrogers.org/episodes/the-first-program/

[xliii] Scooby Doo, Where Are You!/Hassle in the Castle. (1969). episode.

[xliv] Scooby Doo Where Are You!/A Gaggle Of Galloping Ghosts. (1969). episode.

[xlv] *Oops!... I Did it Again (Official HD Video)*. (2010). Retrieved May 26, 2024, from https://www.youtube.com/watch?v=CduA0TULnow.

[xlvi] Britney Spears' "Circus" tour to return to us. (2009, June 10). *San Diego Union-Tribune*.

[xlvii] Crowley, J. (2024, February 9). *Britney Spears dances in a red leather bodysuit & devil's horns in new video*. Hollywood Life. https://hollywoodlife.com/2024/02/09/britney-spears-rocks-red-bodysuit-and-devil-horns-in-dance-video/

xlviii *NRFB Britney Spears Did It Again Video Performance Collection Doll #1731833851*. 24311454 - Online Store. (n.d.). https://huhsaleset.shop/product_details/24311454.html

xlix Amazon.com: Britney Spears Doll "oops!...I did it again" : Toys & games. (n.d.). https://www.amazon.com/Britney-Spears-Doll-Oops-Again/dp/B002NDC3S4

l Amazon.com: Newcotte 2 pcs Halloween pop singer costume metallic bodysuit jumpsuit with fake head microphone for women cosplay : Clothing, shoes & jewelry. (n.d.-b). https://www.amazon.com/Newcotte-Halloween-Metallic-Bodysuit-Microphone/dp/B0C7FX4Y36

li About KIT KAT® Bars | History and FAQs. (n.d.). https://www.hersheyland.com/brands/kit-kat/about.html

lii *TSDR US Serial Number 73044306*. Trademark Status & Document Retrieval. (n.d.). https://tsdr.uspto.gov/#caseNumber=73044306&caseSearchType=US_APPLICATION&caseType=DEFAULT&searchType=statusSearch

liii *TSDR US Serial Number 73716864*. Trademark Status & Document Retrieval. (n.d.). https://tsdr.uspto.gov/#caseNumber=73716864&caseSearchType=US_APPLICATION&caseType=DEFAULT&searchType=statusSearch

liv *TSDR US Serial Number 97939299*. Trademark Status & Document Retrieval. (n.d.). https://tsdr.uspto.gov/#caseNumber=97939299&caseSearchType=US_APPLICATION&caseType=DEFAULT&searchType=statusSearch

lv *What is the mandela effect? 54 Mandela Effect examples*. Good Housekeeping. (n.d.).

https://www.goodhousekeeping.com/life/entertainment/g28438966/mandela-effect-examples/

[lvi] *Cereal box: Fruit Loops (1967).* The Toyroom Repro & Custom Packaging. (n.d.). https://www.the-toyroom.com/products/fruit-loops-1967-cereal-box

[lvii] Pirnia, G. (2017, September 30). *15 solid facts about The Flintstones.* Mental Floss. https://www.mentalfloss.com/article/81462/15-solid-facts-about-flintstones

[lviii] *TSDR US Serial Number 72197335.* Trademark Status & Document Retrieval. (n.d.). https://tsdr.uspto.gov/#caseNumber=72197335&caseSearchType=US_APPLICATION&caseType=DEFAULT&searchType=statusSearch

[lix] *Screen Gems Television.* Screen Gems Television - Closing Logos. (n.d.). https://www.closinglogos.com/page/Screen_Gems_Television#google_vignette

[lx] Straube B. "An overview of the neuro-cognitive processes involved in the encoding, consolidation, and retrieval of true and false memories." *Behavioral and Brain Functions.* 2012;8(1):35. doi:10.1186/1744-9081-8-35

[lxi] Loftus, E. F., & Palmer, J. C. (1974). "Reconstruction of Automobile Destruction: An Example of the Interaction Between Language and Memory." *Journal of Verbal Learning and Verbal Behavior, 13*(5), 585-589.

[lxii] Byrne, P. (2010). *The many worlds of Hugh Everett III: Multiple universes, mutual assured destruction, and the meltdown of a nuclear family.* Oxford University Press.

[lxiii] Vilenkin, A., & Tegmark, M. (2024, February 20). *The case for parallel universes*. Scientific American. https://www.scientificamerican.com/article/multiverse-the-case-for-parallel-universe/

[lxiv] Everett, H. (1957). "Relative State" Formulation of Quantum Mechanics. *Reviews of Modern Physics, 29*(3), 454-462.

[lxv] Reece. (n.d.). *The Berenstein Bears: We are living in our own parallel universe*. The Berenstein Bears: We Are Living in Our Own Parallel Universe. https://www.woodbetween.world/2012/08/the-berenstein-bears-we-are-living-in.html

[lxvi] Bostrom, N. (2003). Are you living in a computer simulation? *Philosophical Quarterly, 53*(211), 243-255. https://doi.org/10.1111/1467-9213.00309

[lxvii] Fenster, M. (2008). *Conspiracy Theories: Secrecy and Power in American Culture*. University of Minnesota Press.

[lxviii] Melley, T. (2000). *Empire of Conspiracy: The Culture of Paranoia in Postwar America*. Cornell University Press.

[lxix] Bergmann, E. (2018). *Conspiracy & Populism: The Politics of Misinformation*. Palgrave Macmillan.

[lxx] Fenster, M. (2008). *Conspiracy Theories: Secrecy and Power in American Culture*. University of Minnesota Press.

[lxxi] Melley, T. (2000). *Empire of Conspiracy: The Culture of Paranoia in Postwar America*. Cornell University Press.

[lxxii] Pennycook, G., & Rand, D. G. (2019). Fighting misinformation on social media using crowdsourced judgments of news source quality. Proceedings of the National Academy of Sciences, *116*(7), 2521-2526.

[lxxiii] Melley, T. (2000). *Empire of Conspiracy: The Culture of Paranoia in Postwar America.* Cornell University Press.

[lxxiv] Allcott, H., & Gentzkow, M. (2017). *Social media and fake news in the 2016 election.* Journal of Economic Perspectives, *31*(2), 211-236.

[lxxv] Janis, I. L. (1982). *Groupthink: Psychological Studies of Policy Decisions and Fiascoes.* Boston: Houghton Mifflin.

[lxxvi] Vosoughi, S., Roy, D., & Aral, S. (2018). *The spread of true and false news online.* Science, *359*(6380), 1146-1151.

[lxxvii] Cohen, P. (2015, May 29). *How the "Chocolate Diet" hoax fooled millions.* CBS News. https://www.cbsnews.com/news/how-the-chocolate-diet-hoax-fooled-millions/

www.ingramcontent.com/pod-product-compliance
Lightning Source LLC
Chambersburg PA
CBHW071831210526
45479CB00001B/93